MODEL-BASED CONTROL OF
A ROBOT MANIPULATOR

MODEL-BASED CONTROL OF A ROBOT MANIPULATOR

Chae H. An
Christopher G. Atkeson
John M. Hollerbach

The MIT Press
Cambridge, Massachusetts
London, England

PUBLISHER'S NOTE

This format is intended to reduce the cost of publishing certain works in book form and to shorten the gap between editorial preparation and final publication. Detailed editing and composition have been avoided by photographing the text of this book directly from the authors' prepared copy.

©1988 Massachusetts Institute of Technology

All rights reserved. No part of this book may be reproduced in any form by any electronic or mechanical means (including photocopying, recording, or information storage and retrieval) without permission in writing from the publisher.

This book was printed and bound in the United States of America.

Library of Congress Cataloging in Publication Data

An, Chae H.
  Model-based control of a robot manipulator.

(The MIT Press series in artificial intelligence)
Bibliography: p.
Includes index.

  1. Robotics. 2. Manipulators (Mechanism)—Automatic control. I. Atkeson, Christopher G. II. Hollerbach, John M. III. Title. IV. Series.
TJ211.A55   1988   629.8'92   87-29892
ISBN 0-262-01102-6

Contents

Series Foreword	xi
Preface	xiii
1 Introduction	**1**
1.1 Arm Trajectory Control	4
1.2 Building Robot Models	8
1.2.1 Motor Modeling	10
1.2.2 Kinematic Calibration	11
1.2.3 Load Estimation	12
1.2.4 Link Estimation	14
1.3 Position Control	16
1.3.1 Independent-Joint PD Control	16
1.3.2 Feedforward Controller	18
1.3.3 Computed Torque Control	19
1.4 Trajectory Learning	20
1.5 Force Control	22
1.5.1 Dynamic Instability in Force Control	23
1.5.2 Cartesian-Based Position Control	24
1.5.3 Cartesian-Based Force Control	27
2 Direct Drive Arms	**31**
2.1 Commercial Manipulators	32
2.2 Direct Drive Arms	37
2.3 MIT Serial Link Direct Drive Arm	41
3 Kinematic Calibration	**49**
3.1 Methods	50
3.2 Identification Procedure	52
3.2.1 Coordinate Representation	52

		3.2.2 Differential Relations	54

		3.2.2	Differential Relations	54
		3.2.3	The Endpoint Variation	56
		3.2.4	Estimating the Endpoint Location	57
		3.2.5	Iterative Estimation Procedure	60
	3.3	Results		61
	3.4	Discussion		63
4	**Estimation of Load Inertial Parameters**			**65**
	4.1	Newton-Euler Formulation		67
		4.1.1	Deriving the Estimation Equations	67
		4.1.2	Estimating the Parameters	71
		4.1.3	Recovering Object and Grip Parameters	72
	4.2	Experimental Results		73
		4.2.1	Estimation on the PUMA Robot	73
		4.2.2	MIT Serial Link Direct Drive Arm	79
	4.3	Discussion		80
		4.3.1	Usefulness of the Algorithm	80
		4.3.2	Sources of Error	81
		4.3.3	Inaccurate Estimates of the Moments of Inertia . .	84
5	**Estimation of Link Inertial Parameters**			**87**
	5.1	Estimation Procedure		90
		5.1.1	Formulation of Newton-Euler Equations	90
		5.1.2	Estimating the Link Parameters	92
	5.2	Experimental Results		94
	5.3	Identifiability of Inertial Parameters		97
	5.4	Discussion		98
6	**Feedforward and Computed Torque Control**			**101**
	6.1	Control Algorithms		102
	6.2	Robot Controller Experiments		103
		6.2.1	Analog/Digital Hybrid Controller	104
		6.2.2	Computed Torque Controller Experiment	107
	6.3	Discussion		110
7	**Model-Based Robot Learning**			**113**
	7.1	Kinematic Learning		115
	7.2	Trajectory Learning		118
	7.3	The Trajectory Learning Algorithm		120
		7.3.1	The Control Problem	120

Contents

    7.3.2 Feedforward Command Initialization 122
    7.3.3 Movement Execution 122
    7.3.4 Feedforward Command Modification 123
7.4 Trajectory Learning Implementation 124
7.5 Using Simplified Models 127
7.6 Trajectory Learning Convergence 129
    7.6.1 Nonlinear Convergence Criteria 129
    7.6.2 Convergence Does Not Guarantee Good Performance. 130
7.7 Discussion 136

8 Dynamic Stability Issues in Force Control   139
8.1 Stability Problems 140
    8.1.1 General Stability Analysis 141
    8.1.2 Example of Unmodeled Dynamics 144
    8.1.3 Experimental Verification of Instability 145
8.2 Compliant Coverings 148
8.3 Adaptation to the Environment Stiffness 149
    8.3.1 Modeling 149
    8.3.2 Least Squares Algorithm 150
    8.3.3 Feasibility 155
8.4 Joint Torque Control 156
    8.4.1 Dominant Pole 157
    8.4.2 One Link Force Control Experiments 159
8.5 Discussion 166

9 Kinematic Stability Issues in Force Control   167
9.1 Intuitive Stability Analysis 170
    9.1.1 Hybrid Control 171
    9.1.2 Stiffness Control 174
9.2 Root Loci, Simulations, and Experiments 175
    9.2.1 Hybrid Control 176
    9.2.2 Resolved Acceleration Force Control 181
    9.2.3 Stiffness Control 184
9.3 Resolved Acceleration Force Control Experiments during Contact 185
    9.3.1 Experimental Setup 188
    9.3.2 Experimental Results 189
9.4 Discussion 191

10 Conclusion 195
  10.1 Assessment of the DDArm 196
  10.2 Further Issues 198

Appendices 201
  Appendix 1: Integral Load Estimation Equations 201
  Appendix 2: Closed Form Dynamics 204
  Appendix 3: Stability Robustness 206
  Appendix 4: Operational Space and Resolved Acceleration . . . 208

References 211

Index 227

Series Foreword

Artificial intelligence is the study of intelligence using the ideas and methods of computation. Unfortunately, a definition of intelligence seems impossible at the moment because intelligence appears to be an amalgam of so many information-processing and information-representation abilities.

Of course psychology, philosophy, linguistics, and related disciplines offer various perspectives and methodologies for studying intelligence. For the most part, however, the theories proposed in these fields are too incomplete and too vaguely stated to be realized in computational terms. Something more is needed, even though valuable ideas, relationships, and constraints can be gleaned from traditional studies of what are, after all, impressive existence proofs that intelligence is in fact possible.

Artificial intelligence offers a new perspective and a new methodology. Its central goal is to make computers intelligent, both to make them more useful and to understand the principles that make intelligence possible. That intelligent computers will be extremely useful is obvious. The more profound point is that artificial intelligence aims to understand intelligence using the ideas and methods of computation, thus offering a radically new and different basis for theory formation. Most of the people doing artificial intelligence believe that these theories will apply to any intelligent information processor, whether biological or solid state.

There are side effects that deserve attention, too. Any program that will successfully model even a small part of intelligence will be inherently massive and complex. Consequently, artificial intelligence continually confronts the limits of computer science technology. The problems encountered have been hard enough and interesting enough to seduce artificial intelligence people into working on them with enthusiasm. It is natural, then, that there has been a steady flow of ideas from artificial intelligence to computer science, and the flow shows no sign of abating.

The purpose of this MIT Press Series in Artificial Intelligence is to provide people in many areas, both professionals and students, with timely,

Series Foreword

detailed information about what is happening on the frontiers in research centers all over the world.

Patrick Henry Winston
Michael Brady

Preface

This book presents a comprehensive approach to the modeling and control of an advanced manipulator, the MIT Serial Link Direct Drive Arm (DDArm). Theoretical insights into position and force control are developed, and are experimentally demonstrated on the DDArm. Self-calibration procedures for link kinematic parameters, link inertial parameters, and load inertial parameters are presented, and the subsequent models are applied to improve control. The integrated presentation on modeling and control can be read as a prescription for robotics.

Our work in this area began in 1982, when Haruhiko Asada joined the MIT Mechanical Engineering Department and embarked on his second generation of direct drive arm designs. The key difference from the first generation design, executed while he was at Carnegie-Mellon University, was the use of brushless rare-earth motors built by the now-defunct ISI Corporation. Asada's second design is well known and is based on a parallel-drive configuration. At the same time as Asada embarked on his second design, he agreed to build for us another direct drive arm based on the ISI motors, but in a true direct drive configuration with the motors mounted at the joints.

The difference in structure of the two manipulators has led to distinct approaches. For the parallel-link direct drive arm, Asada and Kamal Youcef-Toumi thoroughly explored dynamics simplification through geometry and mass balancing to achieve a decoupled and invariant inertia matrix. For the serial link direct drive arm, we have accepted the full nonlinear dynamics of the manipulator, and have accommodated our modeling, algorithms, and control to these dynamics. We believe we have been successful in our approach, even as Asada and Youcef-Toumi have been successful in theirs.

The MIT Artificial Intelligence Laboratory has a long history of building robot devices, including the Minsky/Bennett Arm, the Silver Arm, the MIT Vicarm (designed by Victor Scheinman and after which the

Preface

PUMA is modeled), the Purbrick Arm, and the Utah/MIT Dextrous Hand (with Steve Jacobsen and the University of Utah). Although commercial robots have also been acquired, we have built our own robots to pursue planning and control strategies not implementable on commercial robots.

The impetus for direct drive arm technology in general and for our direct drive arm in particular is the elimination of gears, and the concomitant problems of friction, backlash, gear eccentricity, and joint compliance. The consequences of gears include an inability to control joint torque accurately, to specify a good dynamic model of the robot, and to control endpoint force quickly and accurately. By eliminating gears, it is possible to build an advanced manipulator for research that approaches ideal characteristics of high speed, load, and positional accuracy, and of force controllability.

With our DDArm, we have been able to test and develop advanced control strategies not possible on most other manipulators. Although a great many papers have been written on the theory of robot control, there have been almost no experimental results to validate these theories. Therefore our experimental results on model building, position control, and force control should be of great interest to the robot control community. Although it was an extremely time consuming process, conducting real experiments was important not only in verifying our algorithms but also in discovering unforeseen problems and gaining further insights.

Our general approach of building accurate robot models and then applying the models for high performance control is introduced in Chapter 1. This chapter is also somewhat tutorial, and describes the basics of position and force control. Chapter 2 begins with a general discussion of manipulators and their suitability for research, and then describes the properties of our DDArm. In particular, the control of joint torque is described.

The next three chapters then discuss how accurate kinematic and dynamic models of the robot can be estimated automatically. Chapter 3 discusses how the kinematic parameters, including link lengths and joint offsets, can be automatically calibrated using a motion tracking system. Chapter 4 describes how the inertial parameters of a load that a robot picks up can be identified using a wrist force/torque sensor. Chapter 5 describes a dynamic estimation procedure to obtain the mass, center of mass, and the moments of inertia of the rigid body inertial model of the robot links, based on joint torque sensing.

The models obtained in the earlier chapters 2-5 of this book are

Preface

applied in the later chapters 6-9 for position and force control. It is shown that accurate models are important for all of those different control contexts. Chapter 6 compares the computed torque controller and the feedforward controller, which employ a dynamic model of the robot, to independent-joint PD controllers. Chapter 7 introduces single trajectory learning for position control, where a dynamic model is used in the learning operator to tune the motor commands through repetition to achieve ultra-high trajectory accuracy.

Force control is the subject of chapters 8 and 9. Dynamic instability, caused by hard contact with the environment, is explained in Chapter 8, and a solution based on open-loop joint torque control is presented. Kinematic instability is discussed in Chapter 9, and represents a new finding for certain hybrid position/force controllers. Resolved acceleration force control, equivalent to impedance control and the operational space method, is implemented and compared to other hybrid position/force controllers that do not use a dynamic model.

Notation

Throughout the book, we have used the following notation for mathematical formulas:

matrices (**B**): uppercase bold font
vectors (**b**): lowercase or greek bold font
scalar (b): italic font.

Acknowledgments

Haruhiko Asada originally designed the DDArm, and Harry West and Kamal Youcef-Toumi assisted in the mechanical construction and electronics. Ki Suh interfaced the DDArm to the VAX 11/750 computer, and constructed the racks with motor amplifiers and analog servos. Tom Callahan built the support base for the DDArm. Dave Siegel and Sundar Narasimhan put together the microprocessor system that was used to drive the arm under digital control.

John Griffiths assisted in the implementation of computed torque control. David Bennett conducted the experiments on kinematic calibration using the Watsmart system. Joseph McIntyre collaborated on the trajectory learning work.

Preface

We thank Eric Aboaf and James Korein for reading preliminary versions of the manuscript. We would like to thank the sponsors who contributed to the support of this work: the System Development Foundation, the Defense Advanced Research Projects Agency, the Office of Naval Research, and the National Science Foundation.

MODEL-BASED CONTROL OF
A ROBOT MANIPULATOR

Chapter 1
Introduction

The theme of this book is building and using robot models for control. The intent is to show that model-based control leads to performance superior to control not based on carefully-constructed robot models. Thus the book naturally falls into two parts: identifying a robot model, and applying the model to control.

Building a robot model involves a mathematical formulation of its components:

- motor models for joint torque control,
- kinematic models of link lengths and of locations of joint axes, and
- inertial models of mass, center of mass, and moment of inertia for loads and links.

The parameters of these models need to be measured or estimated by appropriate procedures. The emphasis in this book is on procedures that allow the robot to calibrate itself with minimal human involvement. We fancifully envision the robot waking up in the morning, stretching to calibrate its motors, moving around a bit to identify inertial parameters, and visually observing its end effector in a few positions to identify kinematic parameters.

Controlling a robot involves both position and force variables. The role of models in position control has been extensively elaborated, but less so in force control. One contribution of this book is to show that robot models are as important for force control as they are for position control. In addition, this book addresses the important new area of trajectory learning, where a robot fine tunes one particular trajectory through repetition. Here again the fidelity of the robot model will determine how efficiently the robot can learn.

Despite voluminous publications on the theory of robot control, ranging from PD to nonlinear control, there are almost no experimental results on performance. To be sure, complicated proofs are often given, and occasionally simulations, that supposedly validate an approach. If robot control is to become a scientific endeavor rather than just the pursuit of esoteric mathematics, it must incorporate experimentation to form a critical hypothesize-and-test loop. There simply is no other way to verify convincingly that particular control algorithms work or make a difference, or to guarantee that one is confronting real problems. Experimentation also stimulates discovery, and in fact our results on kinematic instability in force control, discussed later, serendipitously evolved from problems with an actual implementation.

What makes this book relatively unique is its experimental basis: our ideas have been tested and verified on a real robot. The experimental results in this book lend strong validity to our particular control ideas. To a large extent this lack of experimental results is due to the unavailability of high-performance manipulators that are suitable for experimentation. Commercial robots are not suitable for such reasons as high gear ratios, substantial joint friction, and slow movement. A new generation of robots based on direct drive technology is appearing that is promising for research. This book reports on experiments with our *MIT Serial Link Direct Drive Arm* (DDArm), currently one of the few manipulators available anywhere for testing advanced control strategies (Figure 1.1). We expect more of such experimentally suitable manipulators to become available soon.

Specifically, we present a number of advanced implementations in robot control, which either have not been done at all before or which provide one of the few demanding tests of a control strategy.

- Kinematic parameters of the manipulator were automatically calibrated using a motion tracking system.

- Inertial parameters of a grasped object or load were fully identified by a dynamic estimation procedure after a small number of arm movements.

Introduction

Figure 1.1: MIT Serial Link Direct Drive Arm.

- Inertial parameters of each link of the manipulator were identified in a few arm movements.

- Computed torque control and the feedforward controller have been tested for fast arm motion, where dynamic interactions are significant.

- A trajectory learning scheme was implemented that converged in three repetitions close to the repeatability level of the robot.

- A dynamically stable force controller was implemented that is stable against stiff environments but that also has a fast response and steady-state accuracy.

- Resolved acceleration force control, equivalent to the operational space method, was implemented.

In addition to the experimental results, our work has evolved a number of theoretical insights:

- We have shown that the rigid body dynamics of a manipulator can be written as linear equations in terms of the intrinsic inertial parameters, but also that certain of these parameters can only be identified in linear combinations or not at all.

- We have shown that trajectory learning schemes that do not use an accurate dynamic model of the robot converge extremely slowly.

- We have shown, along with others, that force control is essentially high-gain position control, and that the stiffness of the environment multiplies the force control gain to produce a highly underdamped system. Many robot force controllers then become dynamically unstable in the face of noise and unmodeled dynamics.

- We have found a surprising kinematic instability with certain hybrid position/force control strategies. This kind of instability is a new and important result, and occurs only because of the geometric structure of a multi-joint manipulator.

1.1 Arm Trajectory Control

By trajectory control, we intend the most general definition where position and force are simultaneously controlled. To provide a framework for arm trajectory control, a prototypical robot control architecture is outlined in Figure 1.2. The robot control problem consists of two parts:

- *planning*, where a detailed specification of manipulator position and force is given for every instance of time, and

- *control*, i.e., carrying out the plan.

For current robots, planning is usually accomplished by a human programmer, although a goal of this and much other research is to incrementally automate robot programming. Planning can be improved in several ways: the world models used in planning can be refined, better methods for solving the given task can be generated, and the planning methods themselves can be changed. These processes are assumed to be independent of improving execution of a given plan, and will not be addressed in what follows. We will take the plan as fixed, and will focus on execution and making the effector system obey a given plan more closely.

A plan is a complete specification of the motion of the robot in some coordinate system. Often the plan is expressed in *task coordinates,* for example, the Cartesian coordinates of the hand. In Figure 1.2, the variables $\mathbf{x}_d$ represent the desired position of the hand. The trajectory planner may also specify the desired hand velocity $\dot{\mathbf{x}}_d$ and acceleration $\ddot{\mathbf{x}}_d$. Whether the trajectory planner needs to do so depends on the particular controller

Introduction

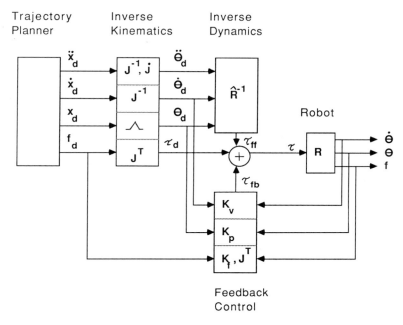

Figure 1.2: A hybrid position/force feedforward controller.

configuration. The trajectory plan may also specify the desired force f_d, when the manipulator is in contact with the environment. The desired force may depend on the position variables, such as in stiffness control or in impedance control. One reason for planning in task coordinates is the ease of partitioning variables into position controlled x_d versus force controlled f_d variables.

In order to execute this plan, task coordinates are converted to *joint coordinates*. To indicate how this is done, it is easiest to first write the *direct kinematics* relationship between the position variables and the corresponding joint variables:

$$x_d = f(\theta_d) \qquad (1.1)$$
$$\dot{x}_d = J\dot{\theta}_d \qquad (1.2)$$
$$\ddot{x}_d = J\ddot{\theta}_d + \dot{J}\dot{\theta}_d \qquad (1.3)$$

The function f is a nonlinear transformation from the joint angles θ_d to the endpoint positions x_d, and depends on the kinematic parameters of the robot. The fidelity of these kinematic parameters determines how accurately the robot may be positioned, and kinematic calibration is one of the basic model-building procedures to be discussed. The direct kinematic velocities and accelerations express linear relationships, with the aid of the Jacobian matrix $J_{ij} = \partial f_i / \partial \theta_j$. Note the slightly more involved expression for $\ddot{\mathbf{x}}_d$.

The adjective *direct* in the expression *direct kinematics* refers to the direction of the transformation, namely from the more internal variable as input (joint variables) to the more external variable as output (endpoint variables). The *inverse kinematics* transformation is required to convert the endpoint trajectory plan to joint variables, by inverting the direct kinematic relationships (1.1)-(1.3):

$$\boldsymbol{\theta}_d = \boldsymbol{\Lambda}(\mathbf{x_d}) \qquad (1.4)$$
$$\dot{\boldsymbol{\theta}}_d = \mathbf{J}^{-1}\dot{\mathbf{x}}_d \qquad (1.5)$$
$$\ddot{\boldsymbol{\theta}}_d = \mathbf{J}^{-1}(\ddot{\mathbf{x}}_d - \dot{\mathbf{J}}\dot{\boldsymbol{\theta}}_d) \qquad (1.6)$$

The nonlinear function $\boldsymbol{\Lambda} = \mathbf{f}^{-1}$ is a one-to-many mapping, and is problematic. For a six degree-of-freedom manipulator, unless the manipulator has the proper kinematic structure, $\boldsymbol{\Lambda}$ cannot be expressed analytically (Tsai and Morgan, 1985). If there are redundancies, then some method must be chosen to resolve the redundancies (Hollerbach and Suh, 1987). In general, we now know that $\boldsymbol{\Lambda}$ cannot be a continuous function covering the whole workspace (Baker and Wampler, 1987).

The expressions (1.5) and (1.6) would not actually be used to evaluate $\dot{\boldsymbol{\theta}}_d$ and $\ddot{\boldsymbol{\theta}}_d$ for efficiency reasons, but rather customized computations would be formed that take advantage of regularities in the robot's kinematic structure (Hollerbach and Sahar, 1983). There is a large volume of literature dealing with the feasibility and computational efficiency of the inverse kinematics transformation, as well as issues of singularities and redundancies. A whole book could be devoted to inverse kinematics, and the topic is not further elaborated here. In our case, the MIT Serial Link Direct Drive Arm only has three degrees of freedom, and inverse kinematics is a relatively simple proposition.

The desired endpoint force $\mathbf{f}_d$ is directly transformed to joint torques $\boldsymbol{\tau}_d$ by the fundamental mechanical relationship:

$$\boldsymbol{\tau}_d = \mathbf{J}^T \mathbf{f}_d \qquad (1.7)$$

This relation is one of *statics* rather than kinematics, but is grouped with kinematics because the transformation involves just the transpose Jacobian matrix $\mathbf{J}^T$. This relation is easy to understand, because the Jacobian matrix is made up of the axes of rotation (Whitney, 1972) and the moment arms about them. In the present case, endpoint forces are reflected to joint torques through the moment arms contained in the Jacobian, and endpoint torques directly sum about the joint axes of rotation. In the positional transformation case, the moment arms produce linear velocities, and the axes of rotation produce angular velocities.

After the inverse kinematics and statics transformation, the next step in making the robot follow a desired trajectory is supplying appropriate commands to the actuators. The simplest approach is *feedback control*, which generates these commands by measuring the difference between where the arm is and where it is supposed to be at any instant in time and using some function (usually a linear function) of this error as the drive signal to the actuators. In Figure 1.2, feedback control is indicated by a box that contains feedback gains $\mathbf{K}_p$ for position error, $\mathbf{K}_v$ for velocity error, and $\mathbf{K}_f$ for force error, to produce an output:

$$\tau_{fb} = \mathbf{K}_p(\boldsymbol{\theta}_d - \boldsymbol{\theta}) + \mathbf{K}_v(\dot{\boldsymbol{\theta}}_d - \dot{\boldsymbol{\theta}}) + \mathbf{J}^T \mathbf{K}_f (\mathbf{f}_d - \mathbf{f}) \qquad (1.8)$$

Note that the force error is computed in task coordinates, and after application of the gain $\mathbf{K}_f$ must be converted to joint torques by the transpose Jacobian $\mathbf{J}^T$.

Feedback control is useful and necessary to compensate for unpredicted disturbances. In particular, when linear feedback control is used alone, the rigid body dynamics of the manipulator are considered as disturbances. These dynamics will cause substantial trajectory errors for faster motions, unless gains in the feedback control are made correspondingly higher. Yet there are practical limits to how high gains can be set, given actuator saturation and stability problems.

To reduce the errors that need to be corrected by feedback, one approach is *feedforward control*, which uses a dynamic model $\hat{R}$ of the robot to predict actuator commands corresponding to a desired motion. This model $\hat{R}$ hopefully represents the actual robot dynamics R fairly accurately, so that the unmodeled dynamics will not cause significant perturbations. Besides the kinematic parameters, the inertial parameters of the links go into the model $\hat{R}$, and their accurate estimation is useful in reducing trajectory errors.

The *direct dynamics* R refers to the transformation from the robot input (joint torques) to the robot output (joint motion); in control terms, R represents the plant dynamics. Again, the adjective *direct* refers to the transformation from the more internal torque variables to the more external joint variables. In Figure 1.2 the calculation of driving torques from a model of the robot and a desired trajectory is then referred to as *inverse dynamics* R^{-1}, obtained by inverting the robot plant R. Since the plant dynamics are not known exactly, only the estimated inverse $\hat{R}^{-1}$ may be used to predict the feedforward torques τ_{ff}.

In practice, there are always unexpected disturbances or modeling errors that make feedforward control imperfect, and a feedback controller is also included to compensate for unpredicted disturbances. Thus the total output τ to the robot is given by:

$$\tau = \tau_{fb} + \tau_{ff} + \tau_d \tag{1.9}$$

Note that the desired torque τ_d based on the planned endpoint force $\mathbf{f}_d$ is also included in this calculation, and can also be thought of as a feedforward command. This control scheme might be termed a *hybrid position/force feedforward controller*, and is only one way to achieve simultaneous control of position and force. Other alternatives will be considered in later sections.

All components of control can be improved using experience. One way of improving the various coordinate transformations involved in robot control is to refine the kinematic model of the robot using measurements from redundant sensing such as vision of the robot tip and joint angle sensing. To improve feedforward control the dynamic model of the robot could be refined. We could also design a better feedback controller using the past history of controller actions.

In this book we will discuss how to build and refine a model of robot dynamics to be used for predicting the appropriate actuator commands to drive the robot (feedforward control). We will discuss how to identify certain types of loads. In addition we will show the role of the robot model as the learning operator during movement practice, i.e., the robot model transforms trajectory following errors into feedforward command corrections.

1.2 Building Robot Models

The first step in any control design is the accurate modeling of the plant to be controlled. In practice, especially with the availability of automatic control design tools, this modeling step may occupy greater than 90% of the control designer's efforts. Hence, for controlling a direct drive arm, accurate modeling of the manipulator is important. The components of

Subsystem	Input	Output
Motor model	Motor input	Joint torques
Kinematic param.	Joint angles	External position sensing
Link inertial param.	Joint torques	Joint movement
Load inertial param.	Joint movement	Wrist force/torque sensing

Table 1.1: Using sensing to decouple system identification problems.

a robot that need to be modeled separately are the motor characteristics, the kinematic parameters, and the inertial parameters.

There are conceptually many different ways in which these components can be determined. For example, kinematic and inertial parameters can be determined from blueprints, and inertial parameters can be determined by taking apart the robot and weighing, counterbalancing, and swinging the pieces. The emphasis in this book is on the robot using input/output relationships between its sensors during movement to calibrate itself, with minimal human involvement. The area of engineering analysis devoted to the problem of characterizing a system from an analysis of input and output data is known as *system identification*. A particular type of system identification that identifies parameters of a known model structure is *parameter estimation*.

An insight that simplifies system identification is to use sensing to decouple different system identification problems. Different identification procedures can then be applied to the different subsystems (Table 1.1). For example, although kinematic and inertial parameters could in principle be identified with the same procedure, in practice it is easier to identify these parameters separately. Kinematic calibration can be accomplished by combining vision and joint angle sensing, while inertial parameter estimation combines joint torque sensing and joint angle sensing.

Widely different rates and types of change in system structure call for different system identification algorithms to track the changes. In the case of an arm with variable loads, arm dynamics are constant or only slowly varying, while load dynamics change rapidly as loads are picked up or put down. Arm identification can be based on a fixed model structure, while a complete load identification system must handle many different model structures, such as rigid loads, flexible loads, and time-varying loads. A wrist force/torque sensor can be used to estimate load parameters independently of the arm dynamics.

Formulation and estimation of the model of the DDArm, consisting of motor modeling, kinematic calibration, load inertial parameter estimation, and link inertial parameter estimation, are presented in Chapters 2-5, and are briefly introduced below.

1.2.1 Motor Modeling

Motor models generally include not only the structure of the motor and amplifier, but also properties of the drive train. Although motor models can be quite complicated, they are in some sense simpler than rigid link dynamic models because motor dynamics are typically confined to a single joint. This reduces motor model identification to a single input/single output modeling problem rather than the more difficult multiple input/multiple output modeling problem.

Because direct drive arms do not have gears, there are no drive train dynamics to model. Hence motor modeling is much easier, because one does not have to contend with friction and backlash. Moreover, the joints are intrinsically stiff, so no extra dynamics are introduced by flexible gear trains. Flexibility in gear trains causes a loss of endpoint precision, and is one of the non-geometric factors that makes kinematic calibration more complicated (Whitney, Lozinski, and Rourke, 1986). Another factor is eccentricity in the gears, which introduces a periodic error in position.

The issue then reduces to how torque can be derived from the motor, given its structure and drive amplifier. The two basic alternatives for an electric motor are current measurement or external torque measurement. Of the two alternatives, current measurement is less accurate, because it is based on a model of the motor: its winding structure, commutation scheme, and magnetic pole location for electromagnetic motors. Sources of error derive from cogging torques and imperfect position measurement. With careful modeling and design, nevertheless, torque accuracies on the order of 1% are now possible in new direct-drive motors (see Chapter 10).

External torque measurement is more accurate, provided that one can design a sensor for the motor axis. The joint structure definitely becomes more complex, especially if a flexible element is introduced to permit sufficient strain to be produced for measurement (Asada, Youcef-Toumi, and Lim, 1984). Additional flexibility may introduce extra dynamics into the system, as well as potential loss of endpoint resolution.

In Chapter 2, the structure of the DDArm is discussed, with particular attention to joint torque control by the motors. Our torque measurement

Introduction

scheme is based on current measurement. Additional points of discussion in that chapter include amplifier nonlinearities, significant inductances, and position inaccuracies of our motors.

1.2.2 Kinematic Calibration

No robot is ever manufactured perfectly. There will be slight variations in kinematic parameters — link lengths that are a little off nominal values or neighboring joint axes that are not quite parallel. The result can be tip inaccuracies on the order of several millimeters for common industrial robots. Kinematic calibration has come to be recognized as a necessary process for any robot, because machining has limits of precision, assembly may be imperfect, and striving for even greater precision is costly.

Other aspects of kinematic calibration, not treated here, include locating the robot with respect to an external reference frame and locating an object within the grasp of the robot. These processes obviously must be carried out whenever a robot is moved or a new object is picked up.

To proceed with kinematic calibration, one formulates how the endpoint position varies with the kinematic parameters. In (1.1), the endpoint position was written only as a function of the joint angles, namely $\mathbf{x} = \mathbf{f}(\boldsymbol{\theta})$. For kinematic calibration, this relation becomes:

$$\mathbf{x} = \mathbf{f}(\boldsymbol{\theta}, \boldsymbol{\alpha}, \mathbf{a}, \mathbf{s}) \tag{1.10}$$

where the Denavit-Hartenberg (1955) parameters $\boldsymbol{\alpha}$, the skew angles between neighboring joint axes, $\mathbf{a}$, the link lengths, and $\mathbf{s}$, the joint offsets, represent the most commonly used kinematic parameters. Taking the first difference,

$$\Delta \mathbf{x} = \frac{\partial \mathbf{f}}{\partial \boldsymbol{\theta}} \Delta \boldsymbol{\theta} + \frac{\partial \mathbf{f}}{\partial \boldsymbol{\alpha}} \Delta \boldsymbol{\alpha} + \frac{\partial \mathbf{f}}{\partial \mathbf{a}} \Delta \mathbf{a} + \frac{\partial \mathbf{f}}{\partial \mathbf{s}} \Delta \mathbf{s} \tag{1.11}$$

The term $\partial \mathbf{f}/\partial \boldsymbol{\theta}$ is just the ordinary Jacobian matrix $\mathbf{J}$, defined in (1.2). The other partial derivatives of $\mathbf{f}$ with respect to the remaining three Denavit-Hartenberg parameters are also different Jacobian matrices.

The difference $\Delta \mathbf{x}$ may be interpreted as the error in position of the endpoint, obtained by subtracting the computed from the measured position. The differences $\Delta \boldsymbol{\theta}$, $\Delta \boldsymbol{\alpha}$, $\Delta \mathbf{a}$, and $\Delta \mathbf{s}$ may be viewed as the corrections to the kinematic parameters. These corrections may be solved for by combining many measurements throughout the workspace and inverting

(1.11):

$$[\Delta\theta \quad \Delta\alpha \quad \Delta a \quad \Delta s] = \begin{bmatrix} \frac{\partial \mathbf{f}}{\partial \theta} & \frac{\partial \mathbf{f}}{\partial \alpha} & \frac{\partial \mathbf{f}}{\partial a} & \frac{\partial \mathbf{f}}{\partial s} \end{bmatrix}^{\dagger} \Delta \mathbf{x} \qquad (1.12)$$

where the † indicates the generalized inverse of the enclosed matrix, giving a least squares solution for the parameter corrections. Since kinematic calibration is a nonlinear estimation problem, one must iterate this procedure by reevaluating the Jacobians at the updated parameter values.

The most difficult aspect of kinematic calibration is obtaining accurate measurements of the endpoint position. Past work in kinematic calibration has involved calibration fixtures with precision points (Hayati and Roston, 1986), special apparatuses (Stone, Sanderson, and Neuman, 1986), or external position measurements with theodolites (Whitney, Lozinski, and Rourke, 1986). Particularly the latter work achieved extremely good results with theodolites, which are manually operated surveying instruments, but one drawback is the extensive human involvement in making the theodolite measurements. Special calibration fixtures also require extensive human involvement, and do not address the issues of locating a robot or an object in the robot's grasp.

Our approach sacrifices to some extent the precision obtained with the above techniques for convenience, so that the robot could in principle calibrate itself. Recently, three-dimensional motion measuring systems have become available that locate points with accuracies on the order of a millimeter and resolutions around a quarter of a millimeter in a volume typical of robot workspaces. We have used one such system, the Watsmart System, which is a commercial opto-electronic apparatus based on triangulation of LED markers. The Watsmart system allows fast tracking of multiple LED markers, and hence the endpoint can be tracked in real time. Kinematic calibration is a straightforward, convenient application of this system, although the accuracies are not comparable to the specialized techniques mentioned above. In Chapter 3 an iterative procedure based on the linearized kinematic equations is presented, along with experimental results using the Watsmart System.

1.2.3 Load Estimation

Since a load is essentially a part of the last link, the knowledge of the inertial parameters of manipulator loads is important for accurate control of manipulators. One alternative is to use object models and precise

Introduction

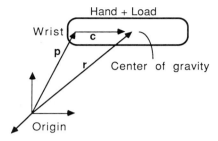

Figure 1.3: Locating the hand plus load center of gravity relative to the wrist and origin.

knowledge of grasping to predict how the load affects the inertial parameters of the last link. A more general alternative would not rely on object models or exact grasping, but would allow a robot to use sensing and an appropriate system identification procedure to identify the load itself.

The general procedure for load estimation begins by writing the Newton-Euler equations for rigid-body motion:

$$\mathbf{f} = m\ddot{\mathbf{r}} \qquad (1.13)$$

$$\mathbf{n} = \mathbf{I}\dot{\boldsymbol{\omega}} + \boldsymbol{\omega} \times \mathbf{I}\boldsymbol{\omega} \qquad (1.14)$$

In Newton's equation, $\mathbf{f}$ is the net force on the body, m is its mass, and $\ddot{\mathbf{r}}$ is the acceleration of the center of gravity. In Euler's equation, $\mathbf{n}$ is the net torque, $\mathbf{I}$ is the inertia and $\boldsymbol{\omega}$ is the angular velocity. Since the position of the center of gravity is not generally known, we decompose $\mathbf{r}$ into the position of the wrist $\mathbf{p}$ plus the position of the load's center of gravity relative to the wrist $\mathbf{c}$ (Figure 1.3). Then

$$\ddot{\mathbf{r}} = \ddot{\mathbf{p}} + \dot{\boldsymbol{\omega}} \times \mathbf{c} + \boldsymbol{\omega} \times (\boldsymbol{\omega} \times \mathbf{c}) \qquad (1.15)$$

Substituting into (1.13),

$$\mathbf{f} = m\ddot{\mathbf{p}} + \dot{\boldsymbol{\omega}} \times m\mathbf{c} + \boldsymbol{\omega} \times (\boldsymbol{\omega} \times m\mathbf{c}) \qquad (1.16)$$

Note that the inertial parameters of mass m, mass moment $m\mathbf{c}$, and inertia $\mathbf{I}$ appear linearly in the Newton-Euler equations (1.14) and (1.16), even though the dynamic equations as a whole are nonlinear. The nonlinearity appears in the kinematic terms, however, which are known as a result of measurement during the motion. By measuring force and torque at the wrist, linear least squares estimation can then be used by combining a number of readings and inverting (1.14)-(1.16):

$$[m \quad m\mathbf{c} \quad \mathbf{I}] = \mathbf{A}(\ddot{\mathbf{p}}, \omega, \dot{\omega})^\dagger [\mathbf{f} \quad \mathbf{n}] \qquad (1.17)$$

where the matrix $\mathbf{A}$ is a function of the kinematic parameters, obtained by rearranging the Newton-Euler equations, and † once again denotes the generalized inverse.

Some previous work in load estimation has specified special test motions, where one joint moves at a time to identify inertia components. The method we present can identify these parameters simultaneously as a result of a single motion. Also, some investigators specify the use of joint torques rather than wrist force/torque sensing, but as mentioned above this couples the dynamics of the arm into the dynamics of the load and makes the estimation less accurate. Finally, most of the previous work has not involved experimentation, so that investigators have not been able to say how well the various load inertial parameters can be estimated.

Our method, presented in Chapter 4, identifies the load inertial parameters after a single motion. Experimental results are presented both on a PUMA 600 robot and on the DDArm.

1.2.4 Link Estimation

The inertial parameters, i.e., the mass, the location of center of mass, and the moments of inertia of each rigid body link of a robot are usually not known even to the manufacturers of the robots. Robots are usually designed to satisfy kinematic specifications, but the inertial parameters are incidental attributes. Since commercial robots are invariably controlled by simple independent-joint PID control, there is no impetus by the manufacturer to determine the inertial parameters accurately since a dynamic model is not used.

As mentioned above, link inertial parameters can be determined experimentally by disassembling the robot, and then weighing the pieces for mass, counterbalancing for center of mass, and swinging for moments of inertia (Armstrong, Khatib, and Burdick, 1986). Besides requiring intensive human involvement, this procedure introduces considerable measurement difficulties. Counterbalanced points have to be referred somehow to the joint axes, while some components of inertia are difficult to determine by pendular motion.

Introduction

Another approach is CAD modeling of the parts, where computerized geometric information can be combined with specific gravities of materials to estimate the inertial parameters. Again, this approach requires intensive human involvement, and is also subject to modeling errors. The approach we present again emphasizes the robot calibrating itself, and will be shown to compare favorably to the other alternatives.

Link inertial parameter estimation is similar to load estimation, in that each link can be viewed as a load to the proximal joints. The inertial parameters again appear linearly in the dynamic equations, even though the multi-link rigid body dynamics are nonlinear. A major difference from load estimation is that there is not a six-axis force/torque sensor at each joint, and only the component of torque about the joint axis can be measured. The inverse dynamics equation can be written as:

$$\tau = \hat{R}^{-1}(\boldsymbol{\theta}, \dot{\boldsymbol{\theta}}, \ddot{\boldsymbol{\theta}}) \qquad (1.18)$$

Since the inertial parameters appear linearly in $\hat{R}^{-1}$, this can be rewritten as:

$$\tau = Q(\boldsymbol{\theta}, \dot{\boldsymbol{\theta}}, \ddot{\boldsymbol{\theta}})[m \ m\mathbf{c} \ \mathbf{I}] \qquad (1.19)$$

where the inertial parameters for all the links have been separated out from $\hat{R}^{-1}$, leaving a function Q of the kinematic parameters only.

Unfortunately, it is not possible to invert (1.19) as before to obtain the link inertial parameters. Aside from the limited sensing at each joint, the proximal links do not undergo general motion because of restricted degrees of freedom. The manifestation of limited sensing and movement in estimation is twofold.

1. Not all parameters can be identified, since they do not influence the joint torques. For example, consider link 1 attached to the robot base by joint 1. Only the link 1 inertia about the first joint is reflected in the joint 1 torque for an arm on a stationary base; all other link 1 inertial parameters cannot be identified.

2. Other inertial parameters can only be determined in linear combinations.

In Chapter 5, we discuss various ways of solving this rank-deficient least squares problem, and present experimental results for our DDArm. One practical aspect is that the parameters that cannot be estimated are unimportant for control, because they do not affect the torques necessary to drive the robot.

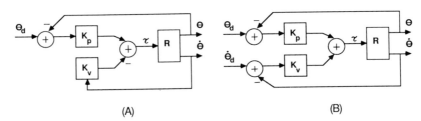

Figure 1.4: Independent-joint PD control with (B) and without (A) velocity reference.

1.3 Position Control

An example of a model-based position/force controller was presented in the beginning of this chapter. This section goes into greater detail about position control, examining the alternatives of how a robot model can be incorporated to improve position control. In a later section we discuss how force control is also improved by using a robot model. We will discuss three basic types of position controllers proposed for robotics:

- independent-joint proportional-derivative (PD) control,
- feedforward control (Liégeois, Fournier, and Aldon, 1980), and
- computed torque control (Paul, 1972).

Independent-joint PD control (Paul, 1981) is by far the most popular feedback controller for robots. As mentioned earlier, independent-joint PD feedback control is distinguished from feedforward control in that the latter uses a model of the robot. Two particular kinds of feedforward control are the *feedforward controller* and the *computed torque controller*, which differ in how the dynamic model is used in conjunction with a feedback loop. Unfortunately, there is a potential confusion in the term *feedforward controller*, which has come to mean in robotics a particular kind of feedforward control. We emphasize again that while much has been written about these and other kinds of robot control, there are too few experimental evaluations of them.

1.3.1 Independent-Joint PD Control

Feedback control can be defined as any control action based on the actual state history of the controlled system. We will restrict the focus to

Introduction

independent-joint PD control. The basic structure of this controller is shown in Figure 1.4A. A reference position $\boldsymbol{\theta}_d$ is compared to an actual position $\boldsymbol{\theta}$, and the difference is multiplied by a position gain $\mathbf{K}_p$ to produce an output to the actuator or plant. To provide stability, a damping term is added to the output based on the actual velocity $\dot{\boldsymbol{\theta}}$ multiplied by a velocity gain $\mathbf{K}_v$:

$$\boldsymbol{\tau} = \mathbf{K}_p(\boldsymbol{\theta}_d - \boldsymbol{\theta}) - \mathbf{K}_v\dot{\boldsymbol{\theta}} \tag{1.20}$$

The output torque τ is applied to the robot, represented by the direct dynamics transformation R, which undergoes a motion $\boldsymbol{\theta}(t)$. Note that feedback control does not include a model of the robot, and is purely error driven. This controller is duplicated for each joint θ of the robot, and the control action at each joint is totally independent of control actions at other joints. Hence the name *independent-joint PD control* is derived.

One feature of this form of PD control is that there is heavy damping during the fastest parts of movement, where it is not particularly needed. To remedy this situation, *PD control with velocity reference* is often proposed (Figure 1.4B), which now requires the trajectory planner to specify the desired velocity $\dot{\boldsymbol{\theta}}_d$ as well as the desired position. Now the damping during the fast parts of movement is reduced to $\mathbf{K}_v(\dot{\boldsymbol{\theta}}_d - \dot{\boldsymbol{\theta}})$.

$$\boldsymbol{\tau} = \mathbf{K}_p(\boldsymbol{\theta}_d - \boldsymbol{\theta}) - \mathbf{K}_v(\dot{\boldsymbol{\theta}}_d - \dot{\boldsymbol{\theta}}) \tag{1.21}$$

Much effort has gone into robot feedback controller design, but there are limits on feedback control. Many proofs of stability for various robot feedback controllers amount to infinite actuator arguments, since it is presumed that actuators do not limit the ability to increase gains to the point where disturbances can be overcome and errors reduced to a desired level. In reality, actuator saturation prevents this easy solution. Moreover, gains cannot be increased to high levels to reduce errors, because of potential instabilities that may arise from modeling error, parameter variations, and measurement or command noise (Åström and Wittenmark, 1984). In general, non-minimum phase elements such as delays and right half plane zeros set limits on maximum feedback gains.

Control of terminal compliance or, more generally, impedance has been proposed as a goal for robotic control. Force control may require limiting feedback gains, if the desired compliance is implemented as a low gain position servo. For non-redundant robots with no terminal force/torque sensing, choosing an impedance specifies the feedback controller completely. The use of force sensors at the interface between the robot and its load or environment may allow differential rejection of modeling errors and external forces, but has not yet been shown to work well.

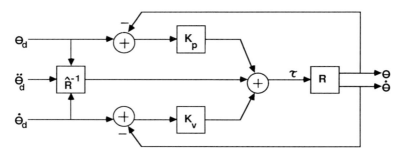

Figure 1.5: Feedforward controller.

1.3.2 Feedforward Controller

The *feedforward controller* (Figure 1.5) predicts the output to the actuators from a dynamic model $\hat{R}^{-1}$ of the robot, based on desired position, velocity, and acceleration. It represents just the position control part of Figure 1.2. Once again, to alleviate a potential source of confusion, it should be repeated that the feedforward controller is a particular kind of the more generic feedforward control.

The trajectory planner must specify not only the desired velocity, but the desired acceleration $\ddot{\boldsymbol{\theta}}_d$ as well. This should be no problem, because the position commands, being internal to the controller, are assumed to have negligible noise and derivative operators can accurately be applied to the command. In parallel with the feedforward computation, there is an independent-joint PD controller with velocity reference. The sum of the feedforward output and the feedback controller output then drives the robot:

$$\boldsymbol{\tau} = \hat{R}^{-1}(\boldsymbol{\theta}_d, \dot{\boldsymbol{\theta}}_d, \ddot{\boldsymbol{\theta}}_d) + \mathbf{K}_p(\boldsymbol{\theta}_d - \boldsymbol{\theta}) + \mathbf{K}_v(\dot{\boldsymbol{\theta}}_d - \dot{\boldsymbol{\theta}}) \tag{1.22}$$

Presumably the feedforward computation has compensated for the dynamics of the robot fairly well, and only small perturbations or unmodeled dynamics remain for the feedback controller to compensate. Hence the gains of the PD controller can be kept low to avoid stability problems.

One issue is the fidelity of the dynamic model $\hat{R}$ of the robot. If the model is not very good, then the feedforward computation can degrade system performance. Our experiments in identification and control with the direct drive arm, however, indicate that this is not a problem.

That the dynamics computation $\hat{R}^{-1}(\boldsymbol{\theta}_d, \dot{\boldsymbol{\theta}}_d, \ddot{\boldsymbol{\theta}}_d)$ in the feedforward controller is done on the basis of the planned trajectory, and hence can

Introduction

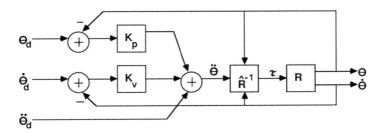

Figure 1.6: Computed torque control.

be done off-line, is an advantage over the computed torque controller discussed next. This may have been an important issue in the past, but it is much less of one today due to increases in computational power of real-time control systems (Narasimhan et al., 1986), and in the efficiency of dynamics computation (Hollerbach and Sahar, 1983). A disadvantage of the feedforward controller is that the PD portion of the controller acts independently of the dynamics and produces perturbations at neighboring joints. That is to say, a corrective torque at one joint perturbs the other joints, whereas ideally the corrective torques would decouple joint interactions. It is to this latter problem that computed torque control is addressed.

1.3.3 Computed Torque Control

In *computed torque control*, the feedback controller sends its output through the dynamic model (Figure 1.6). The feedback control law comprises an independent-joint PD controller with velocity reference, plus the desired acceleration. This yields a corrected acceleration which is then input to the inverse dynamics model:

$$\ddot{\boldsymbol{\theta}}^* = \ddot{\boldsymbol{\theta}}_d + \mathbf{K}_p(\boldsymbol{\theta}_d - \boldsymbol{\theta}) + \mathbf{K}_v(\dot{\boldsymbol{\theta}}_d - \dot{\boldsymbol{\theta}}) \quad (1.23)$$
$$\tau = \hat{R}^{-1}(\boldsymbol{\theta}, \dot{\boldsymbol{\theta}}, \ddot{\boldsymbol{\theta}}^*) \quad (1.24)$$

Note that $\ddot{\boldsymbol{\theta}}^*$, derived from the feedback law, is the nominal rather than the actual acceleration. The feedforward computation $\hat{R}^{-1}(\boldsymbol{\theta}, \dot{\boldsymbol{\theta}}, \ddot{\boldsymbol{\theta}}^*)$ is done on the basis of the actual trajectory, so that the dynamics computation must be on-line.

Computed torque control is a form of control called *non-linearity cancellation*, because if the dynamic model is exact ($\hat{R} = R$), the nonlinear dynamic perturbations are exactly canceled ($\hat{R}^{-1}(R(\tau)) = I(\tau)$, where $I(\tau) = \tau$ is the identity transformation). What is left is a decoupled linear system that can be controlled according to standard techniques. In principle, computed torque control should be more accurate than the feedforward controller, because the action of the feedback controller is decoupled through the dynamics.

As will be seen in Chapter 6, surprisingly the experimental results do not bear out this expectation. Computed torque control and the feedforward controller were about equally accurate in following a trajectory. These results cast into doubt the utility of the ever more sophisticated, or at least complicated, nonlinear feedforward controllers that are being proposed. Once again the importance of experimental testing should be emphasized.

1.4 Trajectory Learning

The implementation of the model building procedures on the DDArm reveals that good models can be identified quickly and are useful for control. Nevertheless, the models used to represent the arm and load dynamics have limited degrees of freedom, and cannot represent the full complexity of the true dynamics. The models do not represent well deviations from the assumed model structure, which are always present to some degree. Thus, on any particular trajectory execution, a rigid body dynamics model will have small errors. Changing the model parameters to fit this trajectory more exactly will degrade performance on other trajectories. What is required is an additional level of modeling that allows representation of fine details of the dynamics.

A solution to this problem, *trajectory learning*, has recently become an important topic in robotics, and arises from the recognition that robots often repeat the same motion over and over again. Hence the possibility arises to tune the output for this single trajectory through repetition to reduce the errors to a very low level. The key issues here are the stability and convergence rate of the iterative process, and how to design the learning operator. One drawback of this approach is that it only produces the appropriate command for a single trajectory. There is little guidance at present as to how to modify that command for similar trajectories.

In trajectory learning, torques are initially generated based on some form of feedforward or feedback controller. The torque profiles are remembered, and refined on a point-by-point basis after each iteration of

Introduction

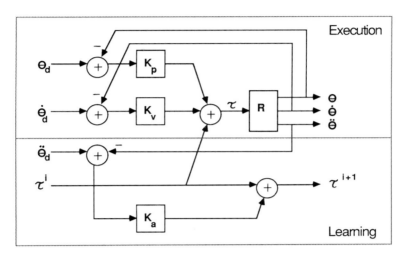

Figure 1.7: Trajectory learning scheme.

the trajectory. The errors during the trajectory are also remembered, and are converted off-line into corrective torques to the torque profiles. One such learning scheme is presented in Figure 1.7, and is similar to one presented in (Arimoto et al., 1984a-c, 1985):

$$\tau = \mathbf{K}_p(\boldsymbol{\theta}_d - \boldsymbol{\theta}) + \mathbf{K}_v(\dot{\boldsymbol{\theta}}_d - \dot{\boldsymbol{\theta}}) + \tau^i \quad (1.25)$$
$$\tau^{i+1} = \tau^i + \mathbf{K}_a(\ddot{\boldsymbol{\theta}}_d - \ddot{\boldsymbol{\theta}}) \quad (1.26)$$

In (1.25) there is a PD position controller with velocity reference, and in addition there is a feedforward torque $\tau^i(t)$ derived from a learning operator after the ith repetition of a movement. At the initial repetition the feedforward torque $\tau^0(t) = 0$, so that the controller is a pure PD position controller the first time. In Figure 1.7, this phase is labeled **Execution**. After the ith repetition, the acceleration error is multiplied by a gain term, and added to the current feedforward torques τ^i in (1.26) to yield the feedforward torques τ^{i+1} for the next movement repetition. This phase is labeled **Learning** in the figure.

This scheme has been shown to converge eventually to a torque profile τ^N that drives the manipulator along a trajectory with small errors. One aspect of this scheme is that a dynamic model $\hat{R}$ of the manipulator is not used. This can be viewed as an advantage in those cases in which a model is not easily obtained. A disadvantage is that the convergence can be very slow.

The relation (1.26) derives a corrective torque by constant scaling of acceleration, but we know from the inverse dynamics R^{-1} that torque and acceleration are not this simply related:

$$\tau = R^{-1}(\boldsymbol{\theta}, \dot{\boldsymbol{\theta}}, \ddot{\boldsymbol{\theta}}) = \mathbf{H}(\boldsymbol{\theta})\ddot{\boldsymbol{\theta}} + \dot{\boldsymbol{\theta}} \cdot \mathbf{C}(\boldsymbol{\theta}) \cdot \dot{\boldsymbol{\theta}} + \mathbf{g}(\boldsymbol{\theta}) \qquad (1.27)$$

where $\mathbf{H}$ is the inertia matrix, $\dot{\boldsymbol{\theta}} \cdot \mathbf{C} \cdot \dot{\boldsymbol{\theta}}$ represents the centripetal and Coriolis torques, and $\mathbf{g}$ represents the gravity torques. This suggests that a better learning scheme would take the manipulator dynamics into account, because one must use an accurate model of the controlled system to make sense of trajectory errors, i.e., convert the errors into corrections to feedforward commands.

Without an accurate model, attempts to improve trajectory performance can actually degrade performance. We have mathematically analyzed the effect of various proposed trajectory learning algorithms such as in Figure 1.7, and can explain why one learning operator works better than another. The most important result is that the convergence rates of the algorithms are determined by the quality of the learning operators used. We can put mathematical bounds on acceptable modeling error for the linear case.

The model used in the trajectory learning work of Chapter 7 is the identified arm model presented in Chapter 5. Thus, starting with only knowledge of system structure, we have demonstrated a system that can build a general model of itself after only three or four movements, and then can learn to execute any particular trajectory to almost the limits of the system repeatability in an additional three or four movements. This work demonstrates the role of knowledge in analyzing past behavior and correcting previous mistakes, and should be compared to other trajectory learning schemes, to table-based schemes to learn arm dynamics, and to much of traditional adaptive control.

1.5 Force Control

Force control is the most general form of trajectory control, because the manipulator is allowed to contact the environment as it executes a trajectory. Instead of just position variables to plan and control, there are now additionally force variables to plan and control. When we use the term *force control,* we mean the simultaneous control of both force and position.

Introduction

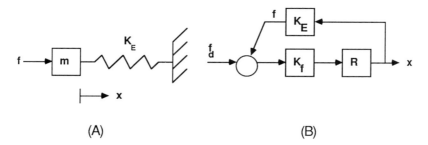

Figure 1.8: (A) A simple model of the robot in contact with the environment. (B) A simple proportional force controller.

Force control is much less well understood than position control, and there have been correspondingly fewer real implementations. As mentioned before, conventional robots are not well suited for implementing force control algorithms since they are essentially positioning devices. Therefore, previous implementations seldom produced satisfactory results (Caine, 1985), and researchers in the past have experienced significant instability problems associated with force controllers (Whitney, 1987). Another goal of this book is to understand some of the stability and performance problems associated with force control, and suggest and demonstrate some remedies to those problems using the DDArm.

In this section, two different aspects of stability in force control are discussed. The first aspect we call *dynamic instability*, which arises when manipulators are in contact with stiff environments. This instability arises whether the robot has single or multiple joints, and is the source of instability in force control most commonly described. The second aspect we call *kinematic instability*, and arises in some Cartesian-based force control schemes only for certain multiple-joint manipulators. We also introduce various types of force controllers, beginning with single-joint controllers and then moving to multiple-joint controllers.

1.5.1 Dynamic Instability in Force Control

Many robot force controllers go unstable during hard contact, such as against metal. The robot chatters uncontrollably, bouncing back and forth against the surface. To illustrate the problem, Figure 1.8A shows a simple model of the robot and its environment. The robot is modeled as a pure mass m, and the environment is modeled as a pure stiffness

K_E. We presume there is a stiff and massless force sensor between the robot and environment that measures the contact force $f = K_E x$, where x is the displacement. The force controller in Figure 1.8B is a simple proportional controller, multiplying the force error $f_d - f$ by a gain K_f, where f_d is the desired force. Hence the applied force f from the force control law is:

$$f = K_f(f_d - K_E x) \tag{1.28}$$

The pertinent feature of this equation is that *force control is essentially high-gain position control*. The stiffness of the environment multiplies the force control gain, yielding a large effective position gain of $K_f K_E$. Systems with such large feedback gains in general exhibit unstable behavior. Sources of instability include unmodeled dynamics, such as flexibility in the manipulator joints or link. Since flexibility is present in all manipulators, the chattering behavior mentioned earlier is manifested in virtually all force controllers.

There are a number of ways in which this dynamic instability can be overcome. One way is to dominate the stiffness of the environment with a soft skin or covering or with a soft spring attaching the force sensor to the robot. Disadvantages with this approach include loss of position resolution and a reduction in the speed of response. The damping in the controller can also be elevated to match the high position gain, but again the response speed would be slowed.

In Chapter 8 we propose a two-part force controller that is dynamically stable but that is still fast and accurate. The fast part of the controller is based on open-loop joint torque control. The stiffness of the environment does not enter into this feedback loop because the external force sensor is not being employed there, and hence the response is always stable. A slower force control loop based on an external force sensor is also used to maintain steady-state accuracy, but the force sensing is low-pass filtered to prevent the environmental stiffness from destabilizing the system. Experimental results are presented with the DDArm.

1.5.2 Cartesian-Based Position Control

As a way of introducing the Cartesian-based force controllers of the next section, it will be helpful to discuss Cartesian-based position controllers at this point. The position controllers discussed earlier (independent-joint PD control, the feedforward controller, and computed torque control) are based on joint coordinates. If the trajectory is initially specified in

Introduction

terms of Cartesian coordinates of the endpoint, then inverse kinematic transformations are required to convert to joint angles.

It is also possible to specify the control law in terms of Cartesian coordinates rather than in terms of joint coordinates. The reason that Cartesian-based position control has been proposed is that the control law is often best cast into the task variables according to which the trajectory is planned. This is particularly true of force control, where variables are partitioned into those that can be controlled for position versus those that are controlled for force. For example, a *Cartesian-based PD position controller* could be defined by:

$$\mathbf{f} = \mathbf{K}_p(\mathbf{x}_d - \mathbf{x}) + \mathbf{K}_v(\dot{\mathbf{x}}_d - \dot{\mathbf{x}}) \quad (1.29)$$

where $\mathbf{f}$ is the endpoint force that the manipulator generates in response to a perturbation. That is to say, the endpoint of the manipulator acts like a spring plus damper.

To evaluate this control law, joint positions and velocities must be converted to endpoint positions and velocities. The endpoint force can be converted into joint torques in several ways. One way is to directly convert to joint torques by the relation $\tau = \mathbf{J}^T\mathbf{f}$ (Figure 1.9):

$$\tau = \mathbf{J}^T(\mathbf{K}_p(\mathbf{x}_d - \mathbf{x}) + \mathbf{K}_v(\dot{\mathbf{x}}_d - \dot{\mathbf{x}})) \quad (1.30)$$

This equation is closely related to Salisbury's stiffness controller, discussed in Chapter 9.

Another way (Figure 1.9B) is to note that $\delta\mathbf{x} = \mathbf{x}_d - \mathbf{x}$ is usually small and can be approximated by the incremental relation $\delta\mathbf{x} = \mathbf{J}\delta\boldsymbol{\theta}$, where $\delta\boldsymbol{\theta} = \boldsymbol{\theta}_d - \boldsymbol{\theta}$. Similarly, $\delta\dot{\mathbf{x}} = \mathbf{J}\delta\dot{\boldsymbol{\theta}}$, where $\delta\dot{\mathbf{x}} = \dot{\mathbf{x}}_d - \dot{\mathbf{x}}$ and $\delta\dot{\boldsymbol{\theta}} = \dot{\boldsymbol{\theta}}_d - \dot{\boldsymbol{\theta}}$. Inverting the Jacobian yields $\delta\boldsymbol{\theta} = \mathbf{J}^{-1}\delta\mathbf{x}$ and $\delta\dot{\boldsymbol{\theta}} = \mathbf{J}^{-1}\delta\dot{\mathbf{x}}$. The gains $\mathbf{K}_p$ and $\mathbf{K}_v$ are now interpreted as joint position and velocity gains, so that

$$\tau = \mathbf{K}_p\mathbf{J}^{-1}(\mathbf{x}_d - \mathbf{x}) + \mathbf{K}_v\mathbf{J}^{-1}(\dot{\mathbf{x}}_d - \dot{\mathbf{x}}) \quad (1.31)$$

This equation is closely related to the hybrid position/force controller of Raibert and Craig, discussed in the next section.

These different implementations of a Cartesian-based PD position controller are pure feedback controllers and do not incorporate a dynamic model of the robot. One could define Cartesian-based position controllers analogous to both the feedforward controller and computed torque control. The position control part of Figure 1.2 would be a hypothetical implementation of a *Cartesian-based feedforward controller*. In actual

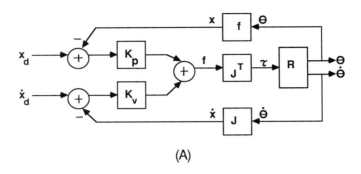

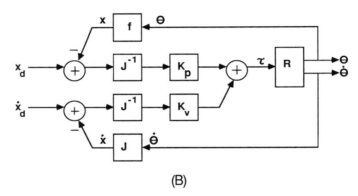

Figure 1.9: (A) Cartesian position control based on the transpose Jacobian matrix. (B) Cartesian position control based on the inverse Jacobian matrix.

implementations and the literature, only the Cartesian-based computed torque controller has been proposed, and is called *resolved acceleration position control* (Luh, Walker, and Paul, 1980b).

Resolved acceleration position control is quite similar to computed torque control, except that the desired trajectory and the feedback law are expressed in terms of task coordinates **x** (Figure 1.10):

$$\ddot{\mathbf{x}}^* = \ddot{\mathbf{x}}_d + \mathbf{K}_p(\mathbf{x}_d - \mathbf{x}) + \mathbf{K}_v(\dot{\mathbf{x}}_d - \dot{\mathbf{x}}) \tag{1.32}$$

Direct kinematic transformations are required to compute the actual endpoint positions and velocities from the joint positions and velocities, and an inverse kinematic transformation is required to convert the nominal

Introduction

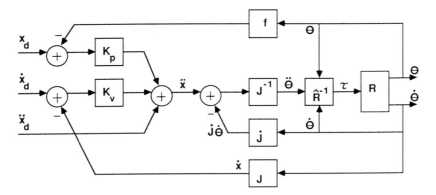

Figure 1.10: Resolved acceleration position control.

endpoint acceleration to a nominal joint acceleration. The nominal joint acceleration is then substituted into the inverse dynamics to yield the joint torques, according to:

$$\ddot{\theta}^* = J^{-1}(\ddot{x}^* - \dot{J}\dot{\theta}) \qquad (1.33)$$
$$\tau = \hat{R}^{-1}(\theta, \dot{\theta}, \ddot{\theta}^*) \qquad (1.34)$$

Resolved acceleration position control is the basis for several Cartesian-based force control schemes discussed next. We have experimentally compared resolved acceleration position control to Cartesian-based PD position control, and found that the former did indeed track the trajectory more accurately.

1.5.3 Cartesian-Based Force Control

When the tip of the manipulator contacts the environment, it will be able to generate positions in certain directions and forces in other directions. Thus the geometry of the environment provides the best coordinate system to partition variables into position-controlled versus force-controlled and to plan the movement (Mason, 1981).

Just as in position control, there are issues of feedback versus feedforward control, and of the role of a model in accurate trajectory tracking. In addition, it turns out there is an issue of instability as well; that is to say, not using a dynamic model can make force control unstable as well as inaccurate. Figures 1.11 and 1.12 illustrate two alternatives of using or not using a model in force control.

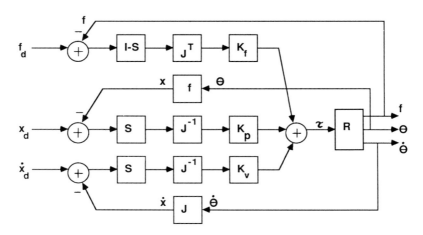

Figure 1.11: Hybrid position/force control.

The controller in Figure 1.11 is known as *hybrid position/force control* (Raibert and Craig, 1981), and does not use a model of the robot. It differs from the Cartesian-based PD position controller based on the inverse Jacobian matrix (Figure 1.9B) only by the inclusion of a force control loop.

$$\tau = \mathbf{K}_p \mathbf{J}^{-1} \mathbf{S}(\mathbf{x}_d - \mathbf{x}) + \mathbf{K}_v \mathbf{J}^{-1} \mathbf{S}(\dot{\mathbf{x}}_d - \dot{\mathbf{x}}) + \mathbf{K}_f \mathbf{J}^T (\mathbf{I} - \mathbf{S})(\mathbf{f}_d - \mathbf{f}) \quad (1.35)$$

The force error $\mathbf{f}_d - \mathbf{f}$ is transformed to joint coordinates by the Jacobian matrix $\mathbf{J}^T$ and then multiplied by the force gain $\mathbf{K}_f$. The force and position feedbacks are summed to provide torques to the robot's joints.

The external variables are presumed to have been partitioned into position-controlled $\mathbf{x}$ versus force-controlled $\mathbf{f}$, and desired trajectories $\mathbf{x}_d$ and $\mathbf{f}_d$ have been specified for each. In the figure, this partitioning is indicated by projection matrices $\mathbf{S}$ and $\mathbf{I} - \mathbf{S}$, where $\mathbf{S}$ selects the variables to be position-controlled (by diagonal elements of 0 and 1), and $\mathbf{I} - \mathbf{S}$ selects the complementary force variables ($\mathbf{I}$ is the identity matrix).

Resolved acceleration force control (Shin and Lee, 1985) is a simple extension to resolved acceleration position control (Figure 1.10), by adding a force loop (Figure 1.12):

$$\tau = \hat{R}^{-1}(\boldsymbol{\theta}, \dot{\boldsymbol{\theta}}, \ddot{\boldsymbol{\theta}}^*) + \mathbf{J}^T (\mathbf{I} - S) \mathbf{K}_f (\mathbf{f}_d - \mathbf{f}) \quad (1.36)$$

A force error $\mathbf{f}_d - \mathbf{f}$ is multiplied first by a force gain $\mathbf{K}_f$ and then by a selection matrix $\mathbf{I} - \mathbf{S}$ to ensure the correct partitioning between position

Introduction

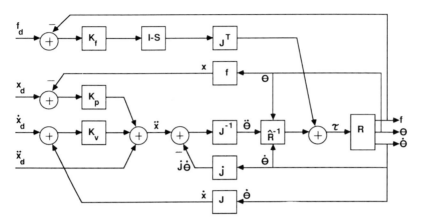

Figure 1.12: Resolved acceleration force control.

variables and force variables. The transformation to joint torques is then accomplished with the transpose Jacobian matrix $\mathbf{J}^T$; note the different order of transformations in force for hybrid position/force control (Figure 1.11). Finally, the outputs from the position and force controllers are summed to produce the joint torques.

Thus the dynamic model is applied to the position control loop, entirely analogous to computed torque control or resolved acceleration position control, and the force control loop adds in separately. Because the transpose Jacobian matrix $\mathbf{J}^T$ directly maps endpoint forces and torques into joint torques, there is no need to map the force control loop through the dynamic model. Resolved acceleration force control is virtually identical to other Cartesian-based force controllers, such as impedance control (Hogan, 1985a-c) and the operational space method (Khatib, 1987), with an advantage perhaps that it is more transparent.

Although hybrid position/force control (Figure 1.11) is quite well known, Chapter 9 presents the surprising result that this control is fundamentally unstable for revolute manipulators. This is a new form of instability that has not been recognized before, and which we call kinematic instability. This instability evidently arises from the interaction of the inverse Jacobian matrix $\mathbf{J}^{-1}$ with the selection matrix $\mathbf{S}$ and the inertia matrix $\mathbf{H}$. Kinematic instability depends on the geometric structure of the manipulator, since polar manipulators do not exhibit this form of instability. The original implementation of Raibert and Craig (1981) was on the Stanford manipulator, which is a polar manipulator, and hence they did not observe any stability problems.

Resolved acceleration force control solves the problem, and is always stable, as long as the dynamic model is reasonably accurate. Note that if the inverse dynamics transformation is modeled as the identity matrix in Figure 1.12, then resolved acceleration force control essentially reduces to hybrid position/force control (Figure 1.11). This poignantly illustrates the importance of a dynamic model in force control as well as in position control, since a substantial modeling error (the identity matrix) makes the controller unstable.

Chapter 9 also presents experiments on resolved acceleration force control with the direct drive arm. The estimated dynamic model from Chapter 5 is used, as well as the two-part dynamically stable force control loop from Chapter 8. The results indicate fast, stable, and accurate force control.

Chapter 2

Direct Drive Arms

As stated earlier, research in manipulator control is at an impasse because of the lack of suitable experimental manipulators. There is a general consensus about what characteristics an advanced manipulator system should have, and most papers on advanced robot control presume some or all of the following characteristics:

- an ideal rigid-body dynamic model of the arm
- fast speed and adequate payload capability
- accurate joint torque control
- accurate position sensing
- accurate velocity sensing
- a force control capability
- adequate bandwidth and accessibility of the robot controller
- adequate computational power for real-time implementation of advanced control algorithms

Unfortunately, there are almost no manipulators that satisfy all or even some of these characteristics. Commercial robots in fact satisfy virtually none of them, and hence cannot serve as experimental testbeds for most theoretical robot control work.

The emergence of direct-drive motor technology has provided a big boost to experimental robotics, because it now appears manipulators can be built that possess the characteristics above. Direct drive robot technology is still in its infancy and beset by a number of problems, and there are only a handful of operational direct drive arms in research labs. Our DDArm is one of the few such arms.

2.1 Commercial Manipulators

Much of commercial robotics is an outgrowth of the machine tool industry, which has viewed robots as basically numerically controlled milling machines. Robot operations are conceived as purely the generation of a sequence of positions, and the structures, servos, and environment are dedicated to this aim.

- To achieve high positional accuracy, robots are designed to be massive and stiff. As a consequence they are slow and relatively weak. Typically, over 90% of actuator output is required merely for lifting the arm bulk itself, leaving very little for the payload.

- Commercial robots are given essentially no contact sensing, since the servo aim is to generate the desired positions regardless of what is happening in the environment.

- The environment must be precisely engineered, including precise jigging and fixturing of parts, because the robot cannot adapt to slight misalignments or variations in parts.

In practice, it is not possible to engineer the environment or even the robot to a sufficiently high precision for position control to perform many tasks successfully. Moreover, many robot operations such as insertions intrinsically require the monitoring and control of force.

Attempts at applying these manipulators for contact operations by instrumenting them with force sensors have not proven very successful. Systems are on the brink of instability and require computationally intensive controllers. There must be a very fast servo response for force control, because contact forces can build up extremely rapidly with essentially no displacement. Usually either the robot or the environment breaks. The massive bulk of the manipulator, coupled with weak actuators and gearing, prevents a fast response of the arm. Insofar as compliance does exist in robots or in fixtured objects, it is accidental and not under control.

Direct Drive Arms   33

One major problem with many commercial manipulators is the use of gears, necessary to amplify the limited torque capabilities of most electric motors. The gears amplify the motor torque by a factor α equal to the gear ratio, allowing the robot designer to use smaller motors. Until recently, increasing the motor size to reduce the gear ratio was not feasible, due to the unfavorable scaling relation between motor torque and combined weight of the motor plus supporting structures. Gears create the following problems.

- *Friction and backlash.* These nonlinear effects are due to preloading, tooth wear, misalignment, and gear eccentricity. They are extremely difficult to model, although parametric models of friction have been attempted (Mukerjee, 1986). Rather than modeling, it seems more appropriate to minimize backlash and friction by mechanical tuning techniques (Dagalakis and Myers, 1985). Friction torques can be so large as to dominate link dynamics. For the PUMA 600 manipulator at the MIT Artificial Intelligence Laboratory, it was measured that the friction terms account for as much as 50% of the motor torques.

- *Joint flexibility.* Particularly for robots with harmonic drives, such as the ASEA robot, the gear elements act like springs and will deflect varying amounts depending on the load and link configuration. Joint flexibility will cause loss of accuracy at the endpoint, particularly complicating kinematic calibration. It also adds undesirable transmission dynamics causing difficulties in designing a wide bandwidth controller (Good, Sweet, and Strobel, 1985).

- *Speed limitations.* All electric motors have a maximum speed at which they can operate, due to back EMF and characteristics of the power amplifiers. Commercial robots often operate near this limit, but the resulting joint speed is not very fast due to the gear reduction. Moreover, the amplifiers impose a slew rate limitation, so that joint acceleration is limited. The end result is that geared robots are relatively slow, and their dynamics are dominated by gravity and friction.

- *Dominance of rotor inertias.* A gear ratio of α multiplies an electric motor's rotor inertia by α^2. Many commercial robots are designed with gear ratios that cause rotor inertia to match or dominate link inertias. For example, a typical gear ratio of 100 : 1 reduces the inertial effects of the links by 10^{-4}. The end result is that the dynamics of commercial robots are well approximated by single joint dynamics, and the nonlinear rigid-body dynamic interactions can be ignored (Goor, 1985a, 1985b; Good, Sweet, and Strobel, 1985). From one standpoint, high rotor inertia is an advantage because it makes control easier: one is dealing with separable and independent joint controllers, and any payload at the end can be ignored.

When these points are taken together, geared robots do not conform to the rigid-body dynamic models hypothesized in most theoretical robot controllers. Dynamic interactions between moving links are insignificant, because (1) rotor inertia dominates link inertia, (2) friction torques dominates inertial torques, and (3) gravity torques dominate inertial torques. Hence robot controllers for commercial robots are designed as parallel single-input, single-output systems.

A more severe problem with commercial robots is the inability to control joint torques, yet virtually all advanced control strategies are predicated on this capability. There are two reasons why joint torque control is difficult to implement on commercial robots.

1. The nonlinear joint dynamics due to gear friction and backlash make the measurement and specification of joint torque very difficult. Motor current cannot then be used to infer joint torque, and the alternative of mounting joint torque sensors at the output side of a gear train is problematical and seldom done (Luh, Fisher, and Paul, 1983).

2. Commercial motor servos are typically position controllers, to which one can only send setpoints. To get at the motor currents, it is often necessary to reverse engineer the controllers. Robot manufacturers are often unwilling to provide specifications to assist in this endeavor for proprietary reasons and also because of safety issues.

Related to the inability to specify joint torques is the unsuitability of commercial manipulators for force control. If working with position controllers, one can only control endpoint force by sending position adjustments to the servos. Position resolution of commercial manipulators is usually not adequate to permit fine force gradations, and the response speed of the position servos is typically too slow. Relying on endpoint force sensing, moreover, brings potential stability problems that comprise one important topic of this book.

High gear ratios also restrict a robot's force control capability. The high rotor inertia presents a large impedance to the environment that is difficult to modulate, and the manipulator will appear stiff and unresponsive. Friction, and in particular static friction or stiction, also reduces the capability of the manipulator to sense and respond to external forces.

As a consequence, commercial robots cannot be used in contact operations. Spray painting and spot welding are typical non-contact applications. Force control is considered a crucial aspect of advanced manipulation such as assembly, and until such a time as suitable manipulators are developed the application of robotics will necessarily be limited.

A final problem with using commercial manipulators is the restrictive computational architecture. As mentioned earlier, users usually must communicate with low-bandwidth position controllers over low-bandwidth ports. Host computers and languages are usually too restrictive: they do not allow sensor-based control strategies to be implemented, and their computational power is too puny. As a result, researchers who obtain commercial robots to investigate control typically play out the following scenario:

> A new robotics research lab purchases a commercial robot to undertake experimental control studies. The lab soon realizes that the programming language and host computer are too limited, and undertakes to rip out the computer system and implement its own. The only catch is that a detailed specification of the servo system is needed, and if the lab is lucky the robot manufacturer will agree to provide this information.
>
> After 2 years the lab has finally managed to get the arm under its own computer control with a custom-made planning and control system. The lab then decides that it is important to include contact sensing, but the remaining servo system and bandwidth are not conducive for incorporating such sensors. After one year, the servo system is patched and the lab is finally able to conduct some experiments in robot control. Ultimately, the manipulator's design is seen to prohibit meaningful control experiments, due to the reasons discussed above. The lab is stymied in its quest to study robot control, and must resort to simulation.

This scenario has been played out time and time again. If we accept the premise that demonstrations by research groups pave the way for industrial applications, then robotics will be held back until such a time as there are suitable experimental manipulators.

To these criticisms, an objection can be made that commercial manipulators are still useful for a number of applications and that controllers should be designed based on the characteristics of existing rather than imagined devices. There is some validity to this objection. All too often in the robot control literature, authors state that a particular control design is intended for commercial robots, without considering that their models do not apply to these robots. Sophisticated nonlinear controllers have even been implemented on these robots, with questionable benefits.

There has been some excellent work on improving the control of existing commercial robots. Goor (1985a-b) designed an independent-joint adaptive controller for the PUMA robot, and improved its ability to move rapidly and accurately. He found that a third-order differential equation best described the motor characteristics. As a result, third-order polynomial trajectories are at least required, because the typical second-order polynomial trajectories supplied to the PUMA represent a step input to the servo system with concomitant oscillation and overshoot. He also found that the servo system was too slow, and that the infrequent position updates contributed to oscillation. Goor found that electronic gains in the motor servos needed to be frequently tuned because of daily variations. Chen (1987) improved Goor's controller by discretization and implementation of a lag-lead compensator. Good, Sweet, and Strobel (1985) demonstrated that the ASEA robot could be well modeled by single-joint dynamics, and that it was necessary to model and compensate for the joint flexibility induced by the harmonic drives.

The work of Goor (1985a, 1985b) and Good, Sweet, and Strobel (1985) indicates that *in robotics it is not an issue of whether to use a model, but of choosing a suitable model.* Their models of commercial robots as single-input, single-output systems are suitable, while the nonlinear models in most published robot control papers are not. The actual characteristics of a physical device must be kept in mind to formulate an accurate model and to be sure one is attacking a real problem.

On the other hand, one should not be satisfied with commercial robots. The performance criteria listed initially in this section are worthy goals for robot design, and it is appropriate to pursue control designs as if such manipulators existed. With the advent of direct drive technology, such manipulators are indeed starting to become available.

2.2 Direct Drive Arms

Direct drive arms are being developed to overcome some of the performance limitations of highly geared robots. Direct drive refers to the elimination of gearing and the direct coupling of links to motor axes. Since there is no gearing to amplify the motor torques, the motors have to be large and powerful to exert large torques. Motor technology has recently produced such motors, making direct drive arms possible. Our direct drive arm uses rare-earth motors with samarium-cobalt magnets, which produce strong magnetic fields and hence high torques (Asada and Youcef-Toumi, 1984). Other direct drive arms use 3-phase variable reluctance motors, which are related to stepping motors and do not use magnets (Curran and Mayer, 1985).

By eliminating gearing, backlash effects are eliminated and joint friction is minimized. Hence joint torques can be controlled and measured much more accurately, because the torque a motor produces is virtually exactly that produced at the joint (Asada, Youcef-Toumi, and Lim, 1984). The joints are also directly backdrivable against the motors, without dissipative losses due to friction.

These characteristics give direct drive arms the potential for serving as good experimental devices for testing advanced arm control strategies. Rigid body dynamics are a good model for these arms, because of the reduction of friction and backlash. Rotor inertias of motors do not overwhelm link dynamics, and the full nonlinear dynamic interactions between moving links are manifested. Hence these manipulators are suitable for testing control algorithms based on link dynamics, and indeed it is necessary to use such sophisticated control algorithms since the full coupled dynamics of the links are reflected directly to the actuators.

Direct drive arms are also well suited towards force control. Since the actuators can be treated as torque sources, they are more suitable for controlling forces and torques at the tip of the manipulator. Conversely, contact forces are directly reflected to the motors without the masking effects of friction. The motors also present a low impedance to the environment, because their inertia has not been multiplied by the squared gear ratio. Hence the robots will be less stiff and more responsive to environmental contact.

Direct drive arm technology was pioneered in the United States by Haruhiko Asada. His first version was developed at the Carnegie-Mellon University (Asada and Kanade, 1981; Asada, Kanade, and Takeyama, 1983). This arm had six degrees of freedom in a true direct drive configuration, but its motors had brushes which caused some problems because of high currents. At MIT, Asada developed a second version with brushless motors and rare-earth magnets from the now-defunct ISI Corporation. Two arms were built with these motors, one in a five-bar-linkage configuration that is not truly direct drive (Asada and Youcef-Toumi, 1984; Asada and Ro, 1985). The emphasis on this arm was a design that produced an invariant and decoupled inertia matrix to simplify control. The second arm, the MIT Serial Link Direct Drive Arm, is at the MIT Artificial Intelligence Laboratory and has motors directly coupled to the joints.

Other direct drive arms have been recently constructed. The CMU Direct-Drive Arm II is a six degree-of-freedom robot that has been constructed with motors similar to the MIT version (Kanade and Schmitz, 1985; Kanade, Khosla, and Tanaka, 1984). The AdeptOne Direct-Drive Robot is a SCARA design that makes use of the 3-phase variable reluctance motors mentioned above (Curran and Mayer, 1985). Kuwahara et al. (1985, 1986) have designed a six degree-of-freedom direct-drive arm, the YEWBOT, that uses two-phase variable reluctance stepping motors with permanent magnets.

There are distinct drawbacks to direct drive arm technology. The large motors make the manipulator large and bulky. Another drawback is the flip side to what makes direct drive arms suitable as advanced research arms: the manipulator dynamics will be sensitive to the changes in loads at the tip of the manipulator, since the load inertial effects are fully reflected to the joints without any reduction through gears.

The most serious problem is related to overheating of the motors, since large current flow through the windings is required to produce necessary large torques necessary to move the large link inertias. In particular, it is difficult for the manipulator to hold itself against gravity without overheating. Designers have pursued alternative manipulator geometries to minimize gravity effects.

- Our direct drive arm replaces the usual shoulder pitch axis with a roll axis in the horizontal plane. This geometry is somewhat unusual, but it does reduce gravitational effects at the shoulder (Figure 1.1).

- The five-bar-linkage design by (Asada and Youcef-Toumi, 1984) removes the elbow motor to the shoulder and reduces the arm mass against which gravity acts. The arm has also been lightened by us-

Direct Drive Arms

Figure 2.1: Semi-direct drive arm based on a six-bar linkage.

ing graphite composite materials (Ramirez, 1984), and redesigned as a six-bar linkage (Figure 2.1).

- The CMU Direct Drive Arm II (Kanade and Schmitz, 1985) replaces the shoulder pitch axis with a yaw axis parallel to the base axis. This makes the first two major joints of the arm into a two-link planar horizontal manipulator (Figure 2.2), and gravity does not directly influence these joints.

- The AdeptOne Robot is a SCARA design with the first two links forming a horizontal planar manipulator not influenced directly by gravity (Curran and Mayer, 1985). Moreover, the second and fourth joints are driven by belts with proximal location of the motors, and are not truly direct drive (Figure 2.3). Again, a more proximal motor location reduces arm inertia.

Figure 2.2: The CMU Direct Drive Arm II is based on a two-link horizontal planar manipulator.

- The YEWBOT (Kuwahara et al., 1985) uses a four-bar linkage to drive the elbow joint, though the shoulder and waist joints are direct drive.

Overheating of the motors is also being combatted with newer designs. The variable reluctance direct drive motor designs, for example, has relatively small power dissipation (Ish-Shalom and Manzer, 1985).

Despite these drawbacks, the ability to model the manipulator accurately by ideal rigid body dynamics makes these manipulators very attractive for control studies and for high performance applications. The absence of a gear train makes the mechanical design of direct drive robots particularly simple, allowing modular designs. It should therefore be easier for laboratories to construct suitable manipulators based on direct drive technology, given also the current proliferation of commercial direct

Direct Drive Arms

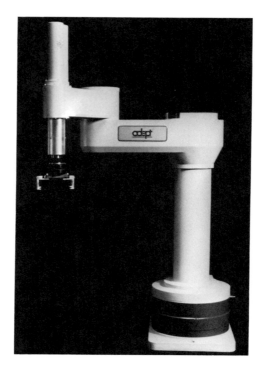

Figure 2.3: The AdeptOne Direct-Drive Robot.

drive motors, and to undertake an advanced experimental investigation of robot control.

2.3 MIT Serial Link Direct Drive Arm

The main experimental device used in this research is the MIT Serial Link Direct Drive Arm (DDArm) (Fig. 1.1). The serial link configuration of this manipulator differentiates this direct drive arm from Asada's parallel-drive arm (Fig. 2.1) also at MIT. The ideal rigid body dynamics is a good model for this arm, uncomplicated by joint friction or backlash.

The coordinate system for this arm is defined in Figure 2.4. A set of inertial parameters is available for the arm (Table 2.1), determined by modeling with a CAD/CAM database (Lee, 1983).

Joint position is measured by a resolver and joint velocity by a tachometer (Figure 2.5). The tachometer is directly attached to the motor

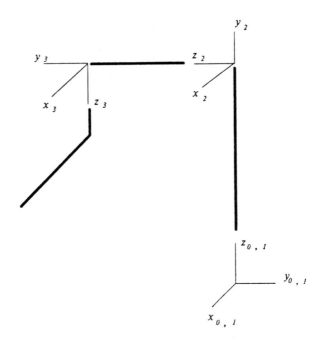

Figure 2.4: The link coordinate system.

Parameters	Link 1	Link 2	Link 3
$m(Kg)$	67.13	53.01	19.67
$mc_x(Kg \cdot m)$	0.0	0.0	0.3108
mc_y	2.432	3.4081	0.0
mc_z	35.8257	16.6505	0.3268
$I_{xx}(Kg \cdot m^2)$	23.1568	7.9088	0.1825
I_{xy}	0.0	0.0	0.0
I_{xz}	-0.3145	0.0	-0.0166
I_{yy}	20.4472	7.6766	0.4560
I_{yz}	-1.2948	-1.5036	0.0
I_{zz}	0.7418	0.6807	0.3900

Table 2.1: CAD-modeled inertial parameters.

Direct Drive Arms

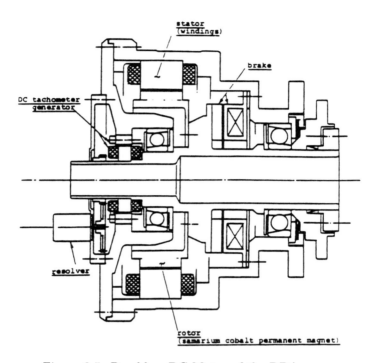

Figure 2.5: Brushless DC Motor of the DDArm.

shaft, while the resolver is geared to the motor shaft. The pancake tachometer output is of high quality (2.2V/rad/s, ripple below 1%). Although the resolver output is digitized with a resolution of 16 bits, in practice the resolution is no better than 12-14 bits. This degradation is directly attributable to the resolver's gears, because of the usual problems of alignment and backlash. As a consequence, the resolver and tachometer outputs are inconsistent at fine resolutions. In this case, we relied on the tachometer output, which we felt was more accurate.

Our current computer system can sample the positions and velocities at up to 1kHz. Joint acceleration is computed by numerically differentiating the velocity signal and low-pass filtering the result using a cutoff frequency of 30Hz. The DDArm is also equipped with a tip force/torque sensor. We used a Barry Wright Company Astek FS6-120A-200 6-axis force/torque sensor which measures all three tip forces and three torques about a point.

Brushless samarium cobalt motors are used to drive the arm (Fig-

Joints	Motor dia. (cm)	Motor mass (kg)	Peak torque (Nm)	Rotor inertia (kg·m^2)	Number of poles	Max. current (Amp) inst.	Max. current (Amp) cont.
1	35	20.39	660	0.181	30	50	15
2 & 3	25	16.5	230	0.0256	18	30	10

Table 2.2: Motor characteristics for the DDArm.

ure 2.5). The motor characteristics of the DDArm are listed in Table 2.2 (Youcef-Toumi, 1985). Under closed-loop control, joint 1 is presently capable of an angular acceleration of 1150 deg/sec^2, joints 2 and 3 in excess of 6000 deg/sec^2. In comparison, joint 1 of the PUMA 600 has a peak acceleration of 130 deg/sec^2 and joint 4 at the wrist 260 deg/sec^2. The motors are driven by 2kHz PWM amplifiers with an output voltage of 350V. With these amplifiers, our motors have an electrical bandwidth of approximately 70 Hz. The bandwidths of the direct drive motors are limited by large motor inductances (Asada and Youcef-Toumi, 1987).

The joint torques are computed by measuring the currents of the 3 phase windings of each motor (Asada, Youcef-Toumi, and Lim, 1984). For the three phase brushless permanent magnet motors of the direct drive arm, the output torque is related to the currents in the windings by:

$$\tau = K_T(I_a \sin\theta + I_b \sin(\theta + 120) + I_c \sin(\theta + 240)) \qquad (2.1)$$

The torque constant K_T for each motor is calibrated statically by measuring the force produced by the motor torque at the end of a known lever arm. The force is measured using the Astek 6-axis force/torque sensor.

Asada, Youcef-Toumi, and Lim (1984) have found that for a motor similar to the motors of our manipulator, the torque versus current relationship was non-linear, especially for small magnitudes of torques, and also varied as a function of the rotor position. The nonlinear effects include deadzone for small torques, cogging, and imperfect commutation circuitry. These nonlinear characteristics were reduced by implementing another torque feedback loop at each joint. Figure 2.6 shows the block diagram of the joint torque servo and also the comparison between the commanded and the measured torques for the third joint of the DDArm with and without the added joint torque feedback. As the plots in Figure 2.6 show, torques can be commanded accurately at each joint. This torque feedback loop was used in the force control experiments of Chapters 8 and 9 only.

Direct Drive Arms

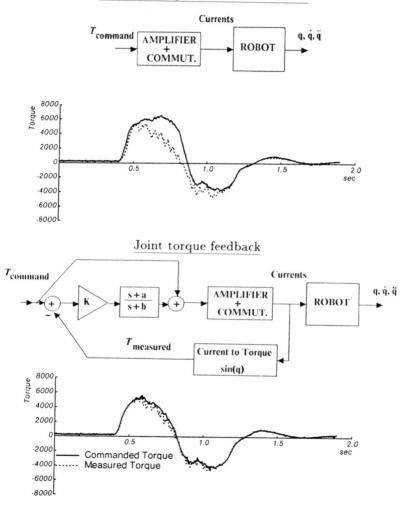

Figure 2.6: With and without explicit joint torque servo.

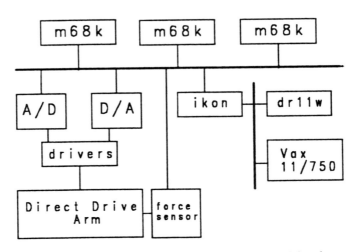

Figure 2.7: Block diagram of the DDArm control hardware.

The robot control system existed in several configurations. Analog controllers could be used to provide constant gain linear feedback control. In addition, a digital computer system could provide feedforward command signals that were summed with the output of the analog controllers. The analog controllers could also be switched off and a pure digital controller could be used.

Our digital control system was based on early versions of the Utah-MIT hand control system (Narasimhan, et al., 1986.). A Multibus cage with several 68000 single board computers and I/O peripherals forms the real-time control hardware. A VAX 11/750, used as the development system, is connected to the real-time Multibus cage through a DMA link. Figure 2.7 shows the hardware configuration of the robot control hardware.

One of the advantages of a direct drive arm is that it can be extremely repeatable. Figure 2.8 shows the position errors on 20 repeated movements of the third joint of the DDArm. Although the absolute errors ($6°$), of a particular controller, are relatively large, they are extremely repeatable. The variation in the errors is on the order of $\pm 0.06°$ during each movement. Most repeatability specifications for commercial robots are for repeatability under static conditions. The high repeatability of this direct drive arm during movement suggests that a model-based control

Direct Drive Arms

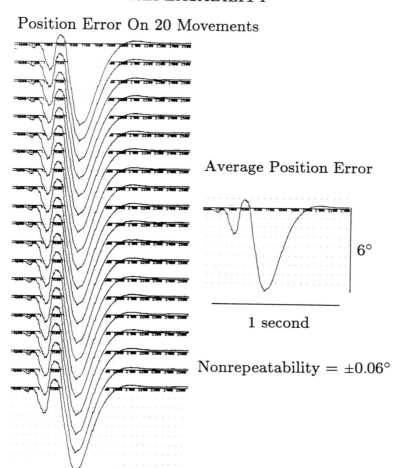

Figure 2.8: Robot repeatability.

system can predict and counteract a large proportion of the trajectory errors of the robot.

Chapter 3
Kinematic Calibration

This chapter investigates the kinematic calibration of a robot manipulator using a motion tracking system. The emphasis in our research is on devising automatic calibration procedures that require minimal human involvement. Another eventual goal of our research is endpoint tracking to compensate for a manipulator on a mobile base. Hence the experimental apparatus should produce accurate estimates in real time of position and orientation. The motion tracking system was selected with these goals, and kinematic calibration falls out as a simple application of this system.

Much of the past experimental work in kinematic calibration has involved labor-intensive apparatuses and procedures. A number of investigators have employed specially machined calibration fixtures with precision points, which require that the operator manually guide an insertion tool (Foulloy and Kelley, 1984; Hayati and Roston, 1986; Veitschegger and Wu, 1987). Other investigators have employed manually operated theodolites (Chao and Yang, 1986; Chen and Chao, 1986; Judd and Knasinski, 1987; Whitney, Lozinski, and Rourke, 1986). Some computer-directed data acquisition systems have also been applied, including a stereo camera system mounted on the end effector (Puskorius and Feldkamp, 1987) and an ultrasonic range sensor (Stone, Sanderson, and Neuman, 1986; Stone and Sanderson, 1987). The stereo camera system, however, has a fairly low accuracy compared to other methods, and the cameras mounted on the end effector obviously interfere with robot operations. The ultrasonic range sensor is fairly accurate and automatic, but the workspace area of operation is quite limited and not suitable for general endpoint tracking.

More sophisticated instrumentation for endpoint tracking has been developed but not yet applied to kinematic calibration. Laser tracking systems appear to be highly accurate (Gilby and Parker, 1982; Lau, Hocken, and Haynes, 1985), with resolutions of 1:100,000 reported. A different class of tracking systems is based on active infrared light-emitting diode markers (IREDs) and cameras with lateral-effect photodiode detectors. The Selspot I, a commercial tracking system produced by Selcom AB of Sweden, was extensively calibrated (Antonsson, 1982) and modified (Dainis and Juberts, 1985) to improve its resolution to 1:4,000. The Selspot II, an improved version, was employed as a robot teaching device, to register object features with a hand-held target and to teach a trajectory (El-Zorkany, Liscano, and Tondu, 1985). A custom-built laboratory apparatus was also used by Ishii et al. (1987) for robot teaching. Although this apparatus appears to be well done, its resolution and sampling rate are not on a par with the commercial systems.

In our experiments, we have also applied a tracking system based on IREDs and lateral-effect photodiode cameras. This Watsmart system is a commercial system produced by Northern Digital Inc. of Waterloo, Ontario, and is roughly comparable to the Selspot II. Individual markers can be sampled at 400 Hz. The resolution of this system is 1:4,000, and its accuracy is about 1 mm at a distance of 2 meters. This accuracy is not as good as that attainable with theodolites, but for our interests accuracy is sacrificed for convenience.

The contribution of the work discussed here, in the face of a large literature in kinematic calibration, is that experimental results in arm calibration are reported for the first time with this kind of motion tracking system. We also feel that the calibration formulation is clearer and more efficient than many of the past efforts. Finally, we present a statistically robust procedure for estimating the endpoint position and orientation from noisy measurements.

3.1 Methods

The DDArm currently has no wrist motions, so the kinematic calibration is limited to three sets of link parameters. We also consider only geometrical factors, since direct drive technology does away with gears and obviates some of the usual complexity in kinematic calibration. The

Kinematic Calibration

Figure 3.1: The DDArm with calibration frame.

non-geometric effects of backlash, gear eccentricity, and joint compliance are simply negligible. We have also not considered base deflection, which is likely to be smaller than the resolution of the measuring system due to the heavy platform construction for the DDArm.

To measure the position and orientation of the last link, a special calibration frame was attached to the third motor (Figure 3.1). This square frame was constructed from welded 1 inch aluminum bars, with outer dimensions of 2 ft. Six infrared light-emitting diodes (IREDs) were attached at 1 ft intervals to the top half of the frame to avoid obscuration, and formed a rectangle of roughly 1 ft by 2 ft. Although three IREDs would have been sufficient to determine the position and orientation, the extra IREDs were included to improve the estimation statistics.

The three-dimensional positions of the IREDs are measured by the Watsmart System. This two-camera system triangulates the markers and calculates their three-dimensional positions after a calibration procedure. Each camera has a special lateral-effect photodiode detector, with a resolution of 1:4000. The cameras are calibrated at the factory with a four-foot plotter moving an IRED through space, to account for lens distortion and electronic effects, and the calibration tables are stored in microprocessors in the system console.

The three-dimensional coordinates of the IREDs are computed by direct linear transformations (DLTs) (Shapiro, 1978), which are set up by a 21 inch calibration cube provided by the manufacturer. The 40 IREDs on this cube were presurveyed at the factory to locate their positions, and this information is stored in the system console microprocessors. The calibration procedure involves placing the cube at several positions in the workspace of interest, and invoking a program supplied by the manufacturer. This camera calibration procedure is convenient and fast: cameras can be placed at will and calibrated within a minute, greatly facilitating different experimental setups. The cameras in this experiment were placed at a distance of about 3 meters from the robot, and the positional accuracy of the calibration was about 1.7 mm (this information is computed by the calibration program based on the DLT fit).

For the kinematic calibration procedure, the calibration frame was placed in 125 different positions, obtained by stepping each joint through 5 increments within a 45° range. These positions and the camera placements were carefully chosen so that all IREDs remained in view. While an exact analysis has not been performed of the optimal placement of the frame for kinematic calibration, the positions were chosen based on a qualitative assessment of the identifiability of the kinematic parameters, and their adequacy was verified by the results.

3.2 Identification Procedure

3.2.1 Coordinate Representation

The Denavit-Hartenberg (1955) link parameters, partially depicted in Figure 3.2 for the DDArm, forms the basis for the calibration procedure. The coordinate origin and axes for link i are found as follows:

- The z_i rotation axis is located at the distal end of link i, and connects links i and $i+1$.

Kinematic Calibration

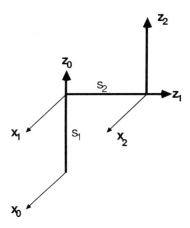

Figure 3.2: The Denavit-Hartenberg link parameters for the DDArm.

- The x_i axis is defined by the common normal between the rotation axes z_{i-1} and z_i.

- The coordinate origin is located at the intersection of the x_i and z_i axes, and is referred to as joint i.

The following four parameters define the location of the link i coordinate system relative to the link $i - 1$ coordinate system:

- θ_i is the joint i angle about the z_{i-1} axis between the x_{i-1} and x_i axes.

- a_i is the distance between z_{i-1} and z_i measured along x_i.

- s_i is the distance between x_{i-1} and x_i measured along z_{i-1}.

- α_i is the angle between z_{i-1} and z_i measured in a righthand sense about x_i.

These link parameters also serve to identify the location of the robot base with respect to the camera coordinate system and of the calibration tool with respect to the robot's last link.

In case of near-parallel neighboring joint axes, the modification suggested in (Hayati, 1983; Hayati and Mirmirani, 1985) could be employed, where an additional skew angle β is substituted for the joint offset s. Six-parameter transformations have been proposed as another way of avoiding the parallel axis problem with the Denavit-Hartenberg parameters (Chao and Yang, 1986; Chen and Chao, 1986; Stone and Sanderson, 1987; Stone, Sanderson and Neuman, 1986; Whitney, Lozinski, and Rourke, 1986), but their added complexity and redundancy do not seem warranted given the Hayati modification. In any event, it was not necessary in our application to use this modification. Our direct drive arm has orthogonal joint axes, and the coordinate systems of the camera relative to the robot and of the calibration tool relative to the last link did not have nearly parallel z axes.

The kinematic calibration procedure must not only identify the relative transformations between the robot's links, but also the transformation between the camera coordinates and the robot base coordinates and between the calibration frame coordinates and the last link coordinates. The camera coordinates are defined by the camera calibration cube, and all IRED measurements are made with respect to these coordinates. The camera calibration cube was placed in the middle of the calibration frame movement. No model of the calibration frame was employed, and the relative location of the IREDs was presumed unknown.

In order to apply the Denavit-Hartenberg parameters to the camera origin without modifying the robot parameters, the camera origin is labeled as the -1 coordinates (Figure 3.3A). The calibration frame attached to the last link is defined by the x_4 and z_4 axes (Figure 3.3B), oriented respectively parallel to the long axis of the IRED rectangle and normal to the rectangle plane. An intermediate coordinate system defined by x_3 and z_3 is required to provide enough parameters for a six-dimensional transformation between the last link and calibration frame.

3.2.2 Differential Relations

The direct kinematic relation between the Denavit-Hartenberg parameters and the endpoint position and orientation **x** is given by:

$$\mathbf{x} = \mathbf{f}(\boldsymbol{\theta}, \boldsymbol{\alpha}, \mathbf{a}, \mathbf{s}) = \mathbf{f}(\boldsymbol{\phi}) \tag{3.1}$$

where $\boldsymbol{\theta}$, $\boldsymbol{\alpha}$, $\mathbf{a}$, and $\mathbf{s}$ are vectors of the Denavit-Hartenberg parameters representing all the coordinate systems (e.g., $\mathbf{s} = (s_0, s_1, \ldots, s_n)$, and $\boldsymbol{\phi} = (\boldsymbol{\theta}, \boldsymbol{\alpha}, \mathbf{a}, \mathbf{s})$ combines all kinematic parameters into one vector.

The calibration is based on iteration of the linearized direct kinematics (3.1), and has been employed by a number of authors (e.g., Puskorius and Feldkamp, 1987; Sugimoto and Okada, 1985; Veitschegger and Wu,

Kinematic Calibration

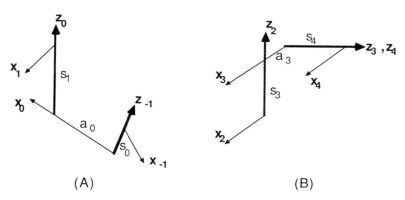

(A)   (B)

Figure 3.3: (A) Coordinate system of the camera relative to the robot base. (B) Coordinate system of the calibration frame relative to the last link.

1987). The first variation $\Delta \mathbf{x}$ of the endpoint location corresponding to variations in the link parameters $\Delta \phi = (\Delta \theta, \Delta \alpha, \Delta a, \Delta s)$ is given by:

$$\Delta \mathbf{x} = \frac{\partial \mathbf{f}}{\partial \theta} \Delta \theta + \frac{\partial \mathbf{f}}{\partial \alpha} \Delta \alpha + \frac{\partial \mathbf{f}}{\partial a} \Delta a + \frac{\partial \mathbf{f}}{\partial s} \Delta s \qquad (3.2)$$

The differential relation may be expressed more compactly as:

$$\Delta \mathbf{x} = \begin{bmatrix} \frac{\partial \mathbf{f}}{\partial \theta} & \frac{\partial \mathbf{f}}{\partial \alpha} & \frac{\partial \mathbf{f}}{\partial a} & \frac{\partial \mathbf{f}}{\partial s} \end{bmatrix} \begin{bmatrix} \Delta \theta \\ \Delta \alpha \\ \Delta a \\ \Delta s \end{bmatrix} \qquad (3.3)$$

$$= \mathbf{C} \Delta \phi \qquad (3.4)$$

Each of the matrices $\mathbf{J}_\theta = \partial \mathbf{f}/\partial \theta$, $\mathbf{J}_\alpha = \partial \mathbf{f}/\partial \alpha$, $\mathbf{J}_a = \partial \mathbf{f}/\partial a$, and $\mathbf{J}_s = \partial \mathbf{f}/\partial s$ represent Jacobians with respect to the particular kinematic parameters. Taking the lead from Whitney (1972), these matrices can be evaluated by simple vector relationships. This contrasts to the normal practice in the kinematic calibration literature where differential homogeneous transformations are used, yielding in our opinion a more cumbersome and less efficient development.

The first Jacobian $\mathbf{J}_\theta$ is just the normal manipulator Jacobian. If the endpoint variation $\Delta \mathbf{x} = (\Delta \mathbf{x}_p, \Delta \mathbf{x}_o)$ is taken to be composed of a positional variation $\Delta \mathbf{x}_p$ followed by an orientational variation $\Delta \mathbf{x}_o$, then the Jacobian's ith column represents the screw coordinates for the ith

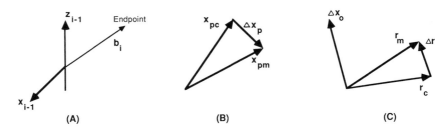

Figure 3.4: (A) Derivation of screw coordinates. (B) Derivation of positional error. (C) Derivation of orientational error.

axis of rotation z_{i-1} (Kumar and Waldron, 1981; Sugimoto and Okada, 1985):

$$\text{col}_i \mathbf{J}_\theta = \begin{bmatrix} \mathbf{b}_i \times \mathbf{z}_{i-1} \\ \mathbf{z}_{i-1} \end{bmatrix} \tag{3.5}$$

where $\mathbf{b}_i$ connects joint i to a reference point on the end effector (Figure 3.4A). The other Jacobians can be found similarly:

$$\text{col}_i \mathbf{J}_\alpha = \begin{bmatrix} \mathbf{b}_i \times \mathbf{x}_i \\ \mathbf{x}_i \end{bmatrix}, \quad \text{col}_i \mathbf{J}_a = \begin{bmatrix} \mathbf{x}_i \\ \mathbf{0} \end{bmatrix}, \quad \text{col}_i \mathbf{J}_s = \begin{bmatrix} \mathbf{z}_{i-1} \\ \mathbf{0} \end{bmatrix} \tag{3.6}$$

3.2.3 The Endpoint Variation

For purposes of kinematic calibration, the endpoint variation $\Delta \mathbf{x}$ is the difference between the measured endpoint location $\mathbf{x}_m$ and the computed endpoint location $\mathbf{x}_c$. This variation can be expressed most conveniently by separating the positional variation from the orientational variation. The positional variation is simply (Figure 3.4B):

$$\Delta \mathbf{x}_p = \mathbf{x}_{pm} - \mathbf{x}_{pc} \tag{3.7}$$

where $\mathbf{x}_{pm}$ is the measured position of a reference point on the end effector and $\mathbf{x}_{pc}$ is the reference point position computed from the link parameters.

The orientational variation $\Delta \mathbf{x}_o$ may be simply found by realizing that it acts like an angular velocity vector (Kumar and Waldron, 1981). Let $\mathbf{r}^*$ be any vector fixed in the end effector coordinate system, and let $\mathbf{A}_m$ and $\mathbf{A}_c$ represent the measured and computed rotation matrices from the

Kinematic Calibration

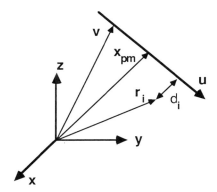

Figure 3.5: Least squares line fit to noisy points.

end effector coordinate system to the global coordinate system. Then (Figure 3.4C)

$$\Delta \mathbf{x}_o \times \mathbf{A}_c \mathbf{r}^* = \mathbf{A}_m \mathbf{r}^* - \mathbf{A}_c \mathbf{r}^* \tag{3.8}$$

This equation may be solved by using the matrix form of the cross product:

$$[\Delta \mathbf{x}_o \times] = (\mathbf{A}_m - \mathbf{A}_c) \mathbf{A}_c^T \tag{3.9}$$

where $\Delta \mathbf{x}_o = (\partial x, \partial y, \partial z)$ and

$$[\Delta \mathbf{x}_o \times] = \begin{bmatrix} 0 & -\partial z & \partial y \\ \partial z & 0 & -\partial x \\ -\partial y & \partial x & 0 \end{bmatrix} \tag{3.10}$$

3.2.4 Estimating the Endpoint Location

A robust estimate of the position and orientation of the end effector can be found from a least squares fit of two straight lines to the set of IREDs. One straight line should lie parallel to the plane formed by the IREDs, while the other straight line should lie normal to this plane. Alternatively, a plane could be fit to the IREDs, but an advantage of the straight line fitting is that the IREDs do not necessarily need to lie in a plane.

Let the equation for a straight line be represented as (Figure 3.5):

$$\mathbf{v} = \mathbf{u} + \mathbf{x}_{pm} = k\hat{\mathbf{u}} + \mathbf{x}_{pm} \tag{3.11}$$

where **v** is any point on the line, **û** is a unit vector pointing in the direction of the line, k is a scalar, $\mathbf{u} = k\hat{\mathbf{u}}$ is any vector pointing along the line, and $\mathbf{x}_{pm}$ is a reference point on the straight line.

For a given IRED position $\mathbf{r}_i$, all three measured coordinates are presumed to have noise. Therefore a least squares fit should be with respect to the normal distance between an IRED position $\mathbf{r}_i$ and the straight line:

$$d_i = \frac{|(\mathbf{r}_i - \mathbf{x}_{pm}) \times \mathbf{u}|}{|\mathbf{u}|} \quad (3.12)$$

The sum of squared errors λ is then:

$$\lambda = \sum_{i=1}^{6} d_i^2 = \sum_{i=1}^{6} \frac{((\mathbf{r}_i - \mathbf{x}_{pm}) \times \mathbf{u})^2}{\mathbf{u}^2} \quad (3.13)$$

where for convenience we write $\mathbf{u}^2 = \mathbf{u} \cdot \mathbf{u}$, etc.

First we estimate the position $\mathbf{x}_{pm}$ of a reference point on the end effector from noisy measurements. Using the relation $a \cdot (b \times c) = (a \times b) \cdot c$, (3.13) may be rewritten as:

$$\lambda = \sum_{i=1}^{6} (\mathbf{r}_i - \mathbf{x}_{pm})^2 - \frac{1}{\mathbf{u}^2} \sum_{i=1}^{6} ((\mathbf{r}_i - \mathbf{x}_{pm}) \cdot \mathbf{u})^2 \quad (3.14)$$

Taking the partial derivative with respect to $\mathbf{x}_{pm}$ and setting to zero yields

$$\frac{\partial \lambda}{\partial \mathbf{x}_{pm}} = \mathbf{0} = -2 \sum_{i=1}^{6} (\mathbf{r}_i - \mathbf{x}_{pm}) + \frac{2}{\mathbf{u}^2} \sum_{i=1}^{6} ((\mathbf{r}_i - \mathbf{x}_{pm}) \cdot \mathbf{u}) \mathbf{u} \quad (3.15)$$

Simplifying,

$$0 = \left(\frac{\mathbf{u}\mathbf{u}^T}{\mathbf{u}^2} - \mathbf{I} \right) \sum_{i=1}^{6} (\mathbf{r}_i - \mathbf{x}_c) \quad (3.16)$$

The general solution to this equation is obtained by equating the second term on the right to zero:

$$\mathbf{x}_{pm} = 1/6 \sum_{i=1}^{6} \mathbf{r}_i \quad (3.17)$$

Thus the most robust estimate of a reference position $\mathbf{x}_{pm}$ is the centroid of the six IREDs forming a rectangle.

Next we estimate the direction **u** of the straight line. This estimation problem differs from ordinary least squares estimation in that it is not

Kinematic Calibration

a linear estimation problem, because **u** appears both in the numerator and denominator. In ordinary least squares, it is presumed that only one of the coordinates has noise, and the projection of error is along that coordinate only. In our case all the coordinates must be presumed to be noisy, and the projection of error along all the coordinates must be taken, i.e., the normal distance to the plane.

Given the reference point $\mathbf{x}_{pm}$ on that line, substitute $\mathbf{x}_i = \mathbf{r}_i - \mathbf{x}_{pm} = (x_i, y_i, z_i)$ into (3.14):

$$\lambda = \sum_{i=1}^{6} \mathbf{x}_i^2 - \frac{1}{\mathbf{u}^2} \sum_{i=1}^{6} (\mathbf{x}_i \cdot \mathbf{u})^2 \tag{3.18}$$

Taking the partial derivative of λ with respect to **u** and equating to zero,

$$\frac{\partial \lambda}{\partial \mathbf{u}} = \mathbf{0} = -\frac{1}{\mathbf{u}^4}(2\mathbf{u}^2 \sum_{i=1}^{6}(\mathbf{x}_i \cdot \mathbf{u})\mathbf{x}_i - 2\mathbf{u}\sum_{i=1}^{6}(\mathbf{x}_i \cdot \mathbf{u})^2) \tag{3.19}$$

Substituting (3.18) and rearranging,

$$((\sum_{i=1}^{6} \mathbf{x}_i^2)\mathbf{I} + \sum_{i=1}^{6} \mathbf{x}_i \mathbf{x}_i^T - \lambda \mathbf{I})\mathbf{u} = 0 \tag{3.20}$$

where **I** is the identity matrix. Let $\mathbf{H} = (\sum_{i=1}^{6} \mathbf{x}_i^2)\mathbf{I} - \sum_{i=1}^{6} \mathbf{x}_i \mathbf{x}_i^T$. Then

$$(\mathbf{H} - \lambda \mathbf{I})\mathbf{u} = 0 \tag{3.21}$$

H represents an ellipsoid of inertia with representative elements $H_{xx} = \sum_{i=1}^{6}(y_i^2 + z_i^2)$, $H_{xy} = -\sum_{i=1}^{6} x_i y_i$, and so on.

Thus the result is an eigenvalue problem (3.21). The cubic eigenvalue equation can be solved analytically. The eigenvectors corresponding to the eigenvalues λ_i represent the principal axes of the ellipsoid of inertia. The smallest eigenvalue λ_1 is the solution for the smallest distance, and so corresponds to an eigenvector $\mathbf{x}_4$ lying in the plane, directed along the long axis of the IRED rectangle. The largest eigenvalue λ_3 corresponds to the normal $\mathbf{z}_4$ to the plane. The vectors $\mathbf{x}_4$ and $\mathbf{z}_4$ are taken to define the coordinate system of the calibration tool, and can be used to derive the rotation matrix $\mathbf{A}_m$ straightforwardly. If the ellipsoid of inertia were symmetric, there would be ambiguity in the eigenvectors, but this is not the case for our IRED rectangular formation.

Joint	s_i (m)	a_i (m)	α_i (rad)	$\Delta\theta_i$ (rad)
0	*	*	*	*
1	*	0.0	-1.571	*
2	-.462	0.0	1.571	3.809
3	*	*	*	*
4	*	N/A	N/A	*

Table 3.1: Initial parameters. The * indicates unknown initial value of a parameter; N/A indicates undefined parameters.

3.2.5 Iterative Estimation Procedure

There are 18 parameters in the vector ϕ that need to be identified for our experimental setup. A number of manipulator poses and measurements are required to determine these parameters robustly. Combine all the error vectors $\Delta\mathbf{x}_i$ and Jacobians into a single equation:

$$\begin{bmatrix} \Delta\mathbf{x}_1 \\ \Delta\mathbf{x}_2 \\ \vdots \\ \Delta\mathbf{x}_m \end{bmatrix} = \begin{bmatrix} \mathbf{C}_1 \\ \mathbf{C}_2 \\ \vdots \\ \mathbf{C}_m \end{bmatrix} \Delta\phi \qquad (3.22)$$

or more compactly,

$$\mathbf{b} = \mathbf{D}\Delta\phi \qquad (3.23)$$

The least squares solution for $\Delta\phi$ is the pseudoinverse of $\mathbf{D}$:

$$\Delta\phi = (\mathbf{D}^T\mathbf{D})^{-1}\mathbf{D}^T\mathbf{b} \qquad (3.24)$$

The updated parameter values ϕ' are then obtained by:

$$\phi' = \phi + \Delta\phi \qquad (3.25)$$

Since this is a nonlinear estimation problem, this procedure is iterated until the variations $\Delta\phi$ approach zero and the parameters ϕ have converged to some stable values. At each iteration, the Jacobians are evaluated with the current parameters.

Joint	s_i (m)	a_i (m)	α_i (rad)	$\Delta\theta_i$ (rad)
0	0.861	-0.099	3.440	0.965
1	0.228	-0.005	-1.587	5.623
2	-0.461	0.011	1.613	3.864
3	-0.542	0.004	-1.552	1.050
4	0.028	N/A	N/A	-0.485

Table 3.2: Calibrated parameters.

3.3 Results

Initial values for the link Denavit-Hartenberg parameters are listed in Table 3.1, as joint numbers 1-2, and were obtained from the nominal design parameters. To begin the iterations, initial values for the camera and calibration tool parameters also had to be chosen, and this was done with manual measurements. The camera coordinates are listed as joint 0 and the calibration tool coordinates as joint 4. Since nothing is attached after joint 4, the a_4 and α_4 parameters are not defined.

The results of the calibration are presented in Table 3.2. The parameters that are intrinsic to the DDArm, rather than to the placement of the camera coordinate system or the calibration frame coordinate system, are all of the index 2 parameters as well as a_1 and α_1. The calibration indicates that neighboring joint axes are not exactly perpendicular ($\alpha_1 = -1.587$ radians and $\alpha_2 = 1.613$ radians). The joint axes also do not intersect ($a_1 = -0.005$ m and $a_2 = 0.011$ m).

The index 0 camera parameters are not particularly meaningful as a point of reference. Since the camera origin defines the $\mathbf{x}_0$ axis, and hence the origin of the base coordinates through the intersection with the $\mathbf{z}_0$ axis, the parameters s_1 and $\Delta\theta_1$ are determined by the camera calibration cube placement.

The calibration frame parameters indicate that the calibration frame was mounted more orthogonally ($\alpha_3 = -1.552$ radians) and closer to intersecting ($a_3 = 0.004$ m) than the manipulator joints. Again, the index 3 and 4 parameters are defined by the placement of the calibration frame, and are not meaningful as a point of comparison.

These parameters converged after 3-6 iterations. A number of different initial conditions were tried, but the iterations always ended up with this result, indicating a global minimum. The iterative least squares method was also compared against the Levenberg-Marquardt algorithm, employed by a number of investigators (Mooring and Tang, 1984). We found that the iterative least squares method generally converged much faster than the Levenberg-Marquardt algorithm. The Levenberg-Marquardt algorithm is more conservative, partially moving along the gradient in small steps. On the other hand, if initial parameters happened to be chosen so that a singularity occurred, then the iterative least squares method could not be used. In this case, the diagonal term in the Levenberg-Marquardt algorithm was essential for convergence. In fact, the Levenberg-Marquardt algorithm is useful to determine initial parameter values for which a good estimate is not available, such as for some of the joint angle offsets. A bad initial estimate quite frequently led to a singularity in the iterative least squares method. After a good initial estimate is available, then it is much better to use the iterative least squares method to make small corrections or to track the robot.

In comparing the calibrated parameters to the initial parameters, one statistical measure is the root mean square (RMS) position and orientation error, defined as:

$$\text{RMS position error} = \sqrt{1/N \sum (\Delta \mathbf{x}_p)^2}$$
$$\text{RMS orientation error} = \sqrt{1/N \sum (\Delta \mathbf{x}_o)^2}$$

where the summation is over the N arm positions the calibration frame was placed in. The position $\Delta \mathbf{x}_p$ and orientation $\Delta \mathbf{x}_o$ error vectors were defined in (3.7) and (3.10). The RMS position error is 10 mm for the initial parameters and 7 mm for the calibrated parameters. The RMS orientation error is 0.027 rad for the initial parameters and 0.021 rad for the calibrated parameters.

Another measure is the percent variance accounted for (VAF), essentially a scaled version of the mean squared error:

$$\text{VAF in position} = 100\% \times \left(1 - \frac{\sum (\Delta \mathbf{x}_p)^2}{\sum (\mathbf{x}_{pm} - \bar{\mathbf{x}}_{pm})^2}\right)$$
$$\text{VAF in orientation} = 100\% \times \left(1 - \frac{\sum (\Delta \mathbf{x}_o)^2}{\sum (\mathbf{x}_{om} - \bar{\mathbf{x}}_{om})^2}\right)$$

For the VAF in position, the normalizing factor is the variance of the measured distance $\mathbf{x}_{pm}$ to the calibration frame centroid, relative to the mean

$\bar{x}_{pm}$ of all centroid positions. For the VAF in orientation, the normalizing factor is defined in terms of the variance of a vector representation of measured orientation of the calibration frame x_{om} relative to its mean over all orientations $\bar{x}_{om}$. When this is done, the VAF is 99.88% for position and 99.92% for orientation for the initial parameters. For the calibrated parameters, the VAF is 99.99% for position and orientation.

An independent test for the two sets of parameters is an external measurement of the deviation of a reference point on the end link from a vertical straight line. A height gauge was used, as suggested in (Whitney, Lozinski, and Rourke, 1986). The reference point was placed manually at several positions along the height gauge, and the corresponding joint angles were read. The endpoint positions were then computed from the joint angles for both the initial parameters and the calibrated parameters, and a straight line was fitted to each set of endpoints. The RMS error of the straight-line fit for the initial parameters was the same, namely 1.2 mm, as that for the calibrated parameters. This number has limited precision, due to problems in locating a reference point on the frame and in aligning this reference point with respect to the height gauge. With this caveat, the height gauge test indicates the two parameter sets are about equally good.

3.4 Discussion

In this chapter we have presented an efficient procedure for kinematic calibration, based on the Denavit-Hartenberg link parameters. The procedure involves iteration of the linearized kinematic equations. One important aspect of this linearization is the use of efficient vector methods to evaluate the various Jacobians and to calculate orientation error. Another contribution is a statistically robust method for determining endpoint position and orientation. The nonlinear least squares fit for position and orientation of the calibration frame was shown to reduce to an eigenvalue problem.

Experimental results in kinematic calibration were obtained for the MIT Serial Link Direct Drive Arm using the Watsmart system, an optoelectronic motion tracking system. The height gauge test did not differentiate between the two parameter sets, since equally good results were obtained. In terms of the RMS error and VAF statistical measures, the calibrated parameters fit the Watsmart data much better than did the initial parameters.

The most probable reason for the discrepancy between the height gauge test and the other two statistical tests is errors in the Watsmart data. The current accuracy of the system is simply not good enough for kinematic calibration yet. We actually feel that the nominal values of the parameters are more correct, and provide a test for the Watsmart system rather than vice versa. Thus the calibration procedure worked well given the experimental measurements, but these measurements have limited accuracy as indicated by the height gauge.

Aside from the current accuracy limitations of the Watsmart system, we feel that the approach that we have offered holds promise for the future. As mentioned earlier, the newest version of the Watsmart system has a resolution of 1:20,000 as compared to the current 1:4,000. This compares more favorably with the experimental laser ranging systems (Gilby and Parker, 1982; Lau, Hocken and Haynes, 1985), which however have not yet been experimentally applied to kinematic calibration. The advantage of our apparatus is convenience, potential accuracy, and automaticity.

We found that the iterative least squares method was sensitive to the initial parameter estimates. A bad estimate was quite likely to lead to a singularity, causing the failure of this method. Although the Levenberg-Marquardt algorithm is much slower, it nevertheless is robust with respect to singularities. Thus a good initial parameter estimate can be found using the Levenberg-Marquardt algorithm, after which the iterative least squares method can be used for updating.

Versions of the Watsmart system are available that calculate three dimensional positions of the markers in real time or that calculate the position and orientation of two bodies at 100 Hz. The ability to track the endpoint location in real time has other uses than just kinematic calibration. We intend to begin investigation of the control of arms on mobile platforms, for which this real-time facility is essential. Endpoint tracking is also useful for various forms of robot teaching (Ishii et al., 1987) and even teleoperator control.

Chapter 4

Estimation of Load Inertial Parameters

This chapter presents a method of estimating all of the inertial parameters of a rigid body load using a wrist force/torque sensor: the mass, the moments of inertia, the location of its center of mass, and the object's orientation relative to a force sensing coordinate system. This procedure has three steps:

1. A Newton-Euler formulation for the rigid body load yields dynamics equations linear in the unknown inertial parameters, when the moment of inertia tensor is expressed about the wrist force sensing coordinate system origin.

2. These inertial parameters are then estimated using a least squares estimation algorithm.

3. The location of the load's center of mass, its orientation, and its principal moments of inertia can be recovered from the sensor referenced estimated parameters.

In principle, there are no restrictions on the movements used to do this load identification, except that if accurate estimation of all the parameters is desired the motion must be sufficiently rich (i.e. occupy more than one orientation with respect to gravity and contain angular accelerations in several different directions) and sometimes special test movements must be used to get accurate estimates of moment of inertia parameters.

There are several applications for this load identification procedure. The accuracy of path following and general control quality of manipulators moving external loads can be improved by incorporating a model of the load into the controller, as the effective inertial parameters of the last link of the manipulator change with the load. The mass, the center of mass, and the moments of inertia constitute a complete set of inertial parameters for an object. In most cases, these parameters form a good description of the object, although they do not uniquely define it. The object may be completely unknown at first and an inertial description of the object may be generated as the robot picks up and moves the object. The robot may also be used in a verification process, in which the desired specification of the object is known and the manipulator examines the object to verify if it is within the tolerances. Recognition, finding the best match of a manipulated object to one among a set of known objects, may also be desired. Finally, the estimated location of the center of mass and the orientation of the principal axis can be used to verify that the manipulator has grasped the object in the desired manner.

A key feature of the algorithm to be presented in this chapter is that the identification takes place while the manipulator is in motion, continuously interpreting wrist force and torque sensory data during a desired manipulation. The algorithm requires measurements of the force and torque due to a load and measurements or estimates of the position, velocity, acceleration, orientation, angular velocity, and angular acceleration of the force sensing coordinate system. It can handle incomplete force and torque measurement by simply eliminating the equations containing missing measurements. The necessary kinematic data can be obtained from the joint angles and, if available, the joint velocities of the manipulator. Also, the inertial parameters of a robot hand can be identified using this algorithm and then the predicted forces and torques due to the hand can be subtracted from the sensed forces and torques, so that only the load is estimated.

This inertial parameter estimation algorithm was implemented using a PUMA 600 robot equipped with an RTI FS-B wrist force/torque sensor, and on the DDArm equipped with a Barry Wright Company Astek FS6-120A-200 6-axis wrist force/torque sensor.

Load Estimation

Previous Work

Paul (1981) described two methods of determining the mass of a load when the manipulator is at rest, one requiring the knowledge of joint torques and the other forces and torques at the wrist. The center of mass and the load moments of inertia were not identified.

Coiffet (1983) utilized joint torque sensing to estimate the mass and center of mass of a load for a robot at rest. Moments of inertia were estimated with special test motions, moving only one axis at a time or applying test torques. Because of the intervening link masses and domination of inertia by the mass moments, joint torque sensing is less accurate than wrist force-torque sensing.

The methods presented in Olsen and Bekey (1985, 1986) and Mukerjee (1984; Mukerjee and Ballard, 1985) are similar to the approach taken in this chapter. Both assumed full force-torque sensing at the wrist to identify the load without special test motions. Unlike the present work, their work did not include experimentally implementing their procedures to verify the correctness of the equations or to determine the accuracy of estimation in the presence of noise and imperfect measurements.

Khalil, Gautier, and Kleinfinger (1986) have suggested a method of identifying the load inertial parameters using joint torque sensing as a part of their link identification procedure. As mentioned before, joint torque sensing is less accurate than wrist force-torque sensing. Also, they did not present any simulation or experimental results.

Recently, Slotine and Li (1987) and Hsu et al. (1987) developed manipulator control algorithms which include on-line adaptation for unknown rigid body load inertial parameters. Their work is based on the earlier observation (Atkeson, An, and Hollerbach, 1985ab; Murkerjee and Ballard, 1985), discussed in the following section, that the dynamics for a rigid body load can be written as a set of linear equations in terms of the unknown inertial parameters.

4.1 Newton-Euler Formulation

4.1.1 Deriving the Estimation Equations

To derive equations for estimating the unknown inertial parameters, the coordinate systems in Figure 4.1 are used to relate different coordinate frames and vectors. **O** is an inertial or base coordinate system, which is fixed in space with gravity pointing along the $-z$ axis. **P** is the force

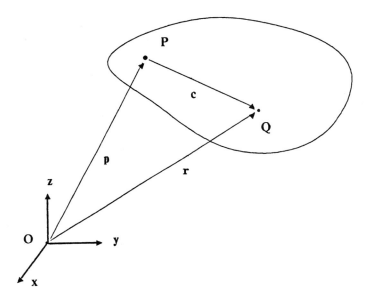

p: position vector from the origin of the base coordinate frame to the origin of the wrist sensor coordinate frame.

r: position vector from the origin of the base coordinate frame to the center of mass of the load.

c: position vector from the origin of the wrist sensor coordinate frame to the center of mass of the load.

Figure 4.1: Coordinate frames.

Load Estimation

reference coordinate system of a wrist force/torque sensor rigidly attached to the load. **Q** represents the principal axis of the rigid body load located at the center of mass. The x axis of **Q** is along the largest principal moment of inertia, and the z axis along the smallest. **Q** is unique up to a reflection in bodies with 3 distinct principal moments of inertia. In the derivation that follows all vectors are initially expressed in the base coordinate system **O**.

The mass, moments of inertia, location of the center of mass, and orientation of the body (a rotation $_{QP}\mathbf{R}$ from the principal axes to the force reference system) are related to the motion of the load and the forces and torques exerted on it by the Newton-Euler equations. The net force $_q\mathbf{f}$ and the net torque $_q\mathbf{n}$ acting on the load at the center of mass are:

$$_q\mathbf{f} = \mathbf{f} + m\mathbf{g} = m\ddot{\mathbf{r}} \qquad (4.1)$$

$$_q\mathbf{n} = \mathbf{n} - \mathbf{c} \times \mathbf{f} = {}_q\mathbf{I}\dot{\boldsymbol{\omega}} + \boldsymbol{\omega} \times ({}_q\mathbf{I}\boldsymbol{\omega}) \qquad (4.2)$$

where:

- **f** = the force exerted by the wrist sensor on the load at the point **p**,
- m = the mass of the load,
- **g** = the gravity vector (**g** = $[0, 0, -9.8\,\text{meters/sec}^2]$),
- $\ddot{\mathbf{r}}$ = the acceleration of the center of mass of the load,
- **n** = the torque exerted by the wrist sensor on the load at the point **p**,
- **c** = the location of the center of mass relative to the force sensing wrist origin **P**,
- $_q\mathbf{I}$ = the moment of inertia tensor about the center of mass,
- $\boldsymbol{\omega}$ = the angular velocity vector, and
- $\dot{\boldsymbol{\omega}}$ = the angular acceleration vector.

To formulate an estimation algorithm, the force and torque measured by the wrist sensor must be expressed in terms of the product of known geometric parameters and the unknown inertial parameters. Although the location of the center of mass and hence its acceleration $\ddot{\mathbf{r}}$ are unknown, $\ddot{\mathbf{r}}$ is related to the the acceleration of the force reference frame $\ddot{\mathbf{p}}$ by

$$\ddot{\mathbf{r}} = \ddot{\mathbf{p}} + \dot{\boldsymbol{\omega}} \times \mathbf{c} + \boldsymbol{\omega} \times (\boldsymbol{\omega} \times \mathbf{c}) \qquad (4.3\,[7.40]^1)$$

---

[1] Equation numbers in brackets refer to equations in Symon, 1971.

Substituting (4.3) into (4.2),

$$\mathbf{f} = m\mathbf{\ddot{p}} - m\mathbf{g} + \dot{\omega} \times m\mathbf{c} + \omega \times (\omega \times m\mathbf{c}) \qquad (4.4)$$

Substituting (4.4) into (4.2),

$$\mathbf{n} = {}_q\mathbf{I}\dot{\omega} + \omega \times ({}_q\mathbf{I}\omega) + m\mathbf{c} \times (\dot{\omega} \times \mathbf{c}) + m\mathbf{c} \times (\omega \times (\omega \times \mathbf{c}))$$
$$+ m\mathbf{c} \times \mathbf{\ddot{p}} - m\mathbf{c} \times \mathbf{g} \qquad (4.5)$$

Although the terms $\mathbf{c} \times (\dot{\omega} \times \mathbf{c})$ and $\mathbf{c} \times (\omega \times (\omega \times \mathbf{c}))$ are quadratic in the unknown location of the center of mass $\mathbf{c}$, these quadratic terms are eliminated by expressing the moment of inertia tensor about the force sensor coordinate origin (${}_p\mathbf{I}$) instead of about the center of mass (${}_q\mathbf{I}$). Equation (4.5) can be rewritten as:

$$\mathbf{n} = {}_q\mathbf{I}\dot{\omega} + \omega \times ({}_q\mathbf{I}\omega) + m[(\mathbf{c}^T\mathbf{c})\mathbf{1} - (\mathbf{c}\mathbf{c}^T)]\dot{\omega}$$
$$+ \omega \times (m[(\mathbf{c}^T\mathbf{c})\mathbf{1} - (\mathbf{c}\mathbf{c}^T)]\omega) + m\mathbf{c} \times \mathbf{\ddot{p}} - m\mathbf{c} \times \mathbf{g}. \qquad (4.6)$$

Using the three dimensional version of the parallel axis theorem

$$_p\mathbf{I} = {}_q\mathbf{I} + m[(\mathbf{c}^T\mathbf{c})\mathbf{1} - (\mathbf{c}\mathbf{c}^T)], \qquad (4.7\ [10.147])$$

(4.6) is then simplified as:

$$\mathbf{n} = {}_p\mathbf{I}\dot{\omega} + \omega \times ({}_p\mathbf{I}\omega) + m\mathbf{c} \times \mathbf{\ddot{p}} - m\mathbf{c} \times \mathbf{g}. \qquad (4.8)$$

(**1** is the 3 dimensional identity matrix). All the vectors are expressed in the wrist sensor coordinate system **P**, so that the quantities $\mathbf{c}$ and ${}_p\mathbf{I}$ are constant. Moreover, the wrist force/torque sensor measures forces and torques directly in the **P** coordinate frame.

In order to formulate the above equations as a system of linear equations, the following notations are used:

$$\omega \times \mathbf{c} = \begin{bmatrix} 0 & -\omega_z & \omega_y \\ \omega_z & 0 & -\omega_x \\ -\omega_y & \omega_x & 0 \end{bmatrix} \begin{bmatrix} c_x \\ c_y \\ c_z \end{bmatrix} \triangleq [\omega \times] \, \mathbf{c} \qquad (4.9)$$

$$\mathbf{I}\omega = \begin{bmatrix} \omega_x & \omega_y & \omega_z & 0 & 0 & 0 \\ 0 & \omega_x & 0 & \omega_y & \omega_z & 0 \\ 0 & 0 & \omega_x & 0 & \omega_y & \omega_z \end{bmatrix} \begin{bmatrix} I_{11} \\ I_{12} \\ I_{13} \\ I_{22} \\ I_{23} \\ I_{33} \end{bmatrix} \triangleq [\bullet\omega] \begin{bmatrix} I_{11} \\ I_{12} \\ I_{13} \\ I_{22} \\ I_{23} \\ I_{33} \end{bmatrix} \qquad (4.10)$$

Load Estimation

where

$$\mathbf{I} = \mathbf{I}^T = \begin{bmatrix} I_{11} & I_{12} & I_{13} \\ I_{12} & I_{22} & I_{23} \\ I_{13} & I_{23} & I_{33} \end{bmatrix} \quad (4.11)$$

Using these expressions, (4.4) and (4.8) can be written as a single matrix equation in the wrist sensor coordinate frame:

$$\begin{bmatrix} f_x \\ f_y \\ f_z \\ n_x \\ n_y \\ n_z \end{bmatrix} = \begin{bmatrix} \ddot{\mathbf{p}} - \mathbf{g} & [\dot{\boldsymbol{\omega}} \times] + [\boldsymbol{\omega} \times][\boldsymbol{\omega} \times] & \mathbf{0} \\ \mathbf{0} & [(\mathbf{g} - \ddot{\mathbf{p}}) \times] & [\bullet \dot{\boldsymbol{\omega}}] + [\boldsymbol{\omega} \times][\bullet \boldsymbol{\omega}] \end{bmatrix} \begin{bmatrix} m \\ mc_x \\ mc_y \\ mc_z \\ I_{11} \\ I_{12} \\ I_{13} \\ I_{22} \\ I_{23} \\ I_{33} \end{bmatrix}$$

(4.12)

or more compactly,

$$\mathbf{w} = \mathbf{A} \boldsymbol{\phi} \quad (4.13)$$

where $\mathbf{w}$ is a 6 element wrench vector combining both the force and torque vectors, $\mathbf{A}$ is a 6×10 matrix and $\boldsymbol{\phi}$ is the vector of the 10 unknown inertial parameters. Note that the center of mass cannot be determined directly, but only as the mass moment $m\mathbf{c}$. But since the mass m is separately determined, its contribution can be factored from the mass moment later. It should be emphasized that the equation (4.13) for the dynamics of a rigid body load is linear in terms of the unknown inertial parameters $\boldsymbol{\phi}$.

4.1.2 Estimating the Parameters

The quantities inside the $\mathbf{A}$ matrix are computed by direct kinematics computation (Luh, Walker, and Paul, 1980a) from the measured joint angles. The elements of the $\mathbf{w}$ vector are measured directly by the wrist force sensor. Since (4.13) represents 6 equations and 10 unknowns, at least two data points are necessary to solve for the $\boldsymbol{\phi}$ vector, i.e. the force and the position data sampled at two different configurations of the manipulator. For robust estimates in the presence of noise, a larger number of data points must be used. Each data point adds 6 more equations, whereas the number of unknowns, the elements of $\boldsymbol{\phi}$, remain constant.

This can be represented by augmenting **w** and **A** as:

$$\mathbf{A} = \begin{bmatrix} \boldsymbol{A}[1] \\ \cdot \\ \cdot \\ \boldsymbol{A}[n] \end{bmatrix}, \quad \mathbf{w} = \begin{bmatrix} \boldsymbol{w}[1] \\ \cdot \\ \cdot \\ \boldsymbol{w}[n] \end{bmatrix}, \quad n = \text{number of data points} \quad (4.14)$$

where each $\boldsymbol{A}[i]$ and $\boldsymbol{w}[i]$ are matrix and vector quantities described in (4.12). Formulated this way, any linear parameter estimation algorithm can be used to identify the $\boldsymbol{\phi}$ vector. A simple and popular method is the least squares method. The estimate for $\boldsymbol{\phi}$ is given by:

$$\hat{\boldsymbol{\phi}} = (\mathbf{A}^T \mathbf{A})^{-1} \mathbf{A}^T \mathbf{w} \quad (4.15)$$

Equation (4.15) can also be formulated in a recursive form (Ljung and Soderstrom, 1983) for on-line estimation.

4.1.3 Recovering Object and Grip Parameters

The estimated inertial parameters (m, $m\mathbf{c}$, $_p\mathbf{I}$) are adequate for control, but for object recognition and verification it is also necessary to obtain the principal moments of inertia I_1, I_2, I_3, the location of the center of mass **c**, and the orientation $_{QP}\mathbf{R}$ of **Q** with respect to **P**.

The parallel axis theorem is used to compute the inertia terms translated to the center of mass of the load.

$$\hat{\mathbf{c}} = \frac{\widehat{m\mathbf{c}}}{\hat{m}} \quad (4.16)$$

$$_q\hat{\mathbf{I}} = \,_p\hat{\mathbf{I}} - \hat{m}[(\hat{\mathbf{c}}^T\hat{\mathbf{c}})\mathbf{1} - (\hat{\mathbf{c}}\hat{\mathbf{c}}^T)] \quad (4.17)$$

The principal moments are obtained by diagonalizing $_q\hat{\mathbf{I}}$.

$$_q\hat{\mathbf{I}} = \,_{QP}\hat{\mathbf{R}} \begin{bmatrix} \hat{I}_1 & 0 & 0 \\ 0 & \hat{I}_2 & 0 \\ 0 & 0 & \hat{I}_3 \end{bmatrix} _{QP}\hat{\mathbf{R}}^T \quad (4.18)$$

This diagonalization can always be achieved since $_q\hat{\mathbf{I}}$ is symmetric; but when two or more principal moments are equal, the rotation matrix $_{QP}\mathbf{R}$ is no longer unique.

Load Estimation 73

Figure 4.2: Puma with a test load.

4.2 Experimental Results

4.2.1 Estimation on the PUMA Robot

The inertial parameter estimation algorithm was initially implemented on a PUMA 600 robot equipped with an RTI FS-B wrist force/torque sensor (Figure 4.2), which measures all six forces and torques. The PUMA 600 has encoders at each joint to measure joint angles, but no tachometers. Thus, to obtain the joint velocities and accelerations, the joint angles are differentiated and double-differentiated, respectively, by a digital differentiating filter (Figure 4.3). The cutoff frequency of 33 Hz for the filter was determined empirically to produce the best results. Both the encoder and the wrist sensor data were initially sampled at 1000 Hz. It was later determined that a sampling rate of 200 Hz was sufficient, and the data were resampled at the lower rate to reduce processing time. A least squares identification algorithm was implemented as an off-line computation, but an on-line implementation would have been straightforward.

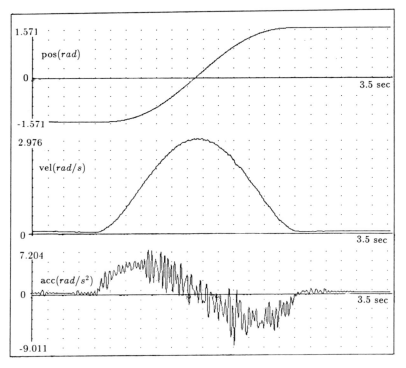

Figure 4.3: Measured angle θ, calculated angular velocity $\dot{\theta}$, and calculated angular acceleration $\ddot{\theta}$ for joint 4.

Load Estimation

Parameters	Actual Values	Static Estimates	Dynamic Estimates
Mass (kg)	1.106	1.103	1.067
Change in $c_y (m)$	0.037	0.037	0.039
Mass (kg)	1.106	1.107	1.084
Change in $c_y (m)$	-0.043	-0.043	-0.042
Mass (kg)	1.106	1.100	1.073
Change in $c_y (m)$	-0.021	-0.020	-0.021
Mass (kg)	1.106	1.099	1.074
Change in $c_y (m)$	0.018	0.018	0.020

Table 4.1: Mass and the center of mass estimates.

Static Estimation Using the PUMA

To test the calibration of the force sensor and the kinematics of the PUMA arm a static identification was performed. The forces and torques are now due only to the gravity acting on the load, and equations (4.4) and (4.8) simplify to

$$\mathbf{f} = -m\mathbf{g} \qquad (4.19)$$
$$\mathbf{n} = -m\mathbf{c} \times \mathbf{g} \qquad (4.20)$$

As seen in (4.19) and (4.20), only the mass and the center of mass can be identified while the manipulator is stationary.

To avoid needing to determine the gripper geometric parameters, the center of mass estimates are evaluated by the estimates of the changes in the center of mass as the load is moved along the y-axis from the reference position by known amounts. The results of estimation are shown in the third column of Table 4.1 for an aluminum block ($2 \times 2 \times 6\,in.$) with a mass of $1.106\,Kg$. Only the changes in c_y are shown in Table 4.1; the estimates of c_x and c_z remained within $1\,mm$ of the reference values ($c_x = 1\,mm$ and $c_z = 47\,mm$). Each set of estimates was computed from 6 sets of data, i.e. data taken at 6 different positions and orientations of the manipulator, where each data point is averaged over 1000 samples to minimize the effects of noise. The results show that in the static case the mass of the load can be estimated to within $10g$ of the actual mass. The center of mass can be estimated to within $1mm$ of the actual values for this load.

Static load estimation only tests the force sensor calibration and the position measurement capabilities of the robot on which the sensor is mounted. In order to assess the effects of the dynamic capabilities of the robot on load estimation and to be able to estimate the moments of inertia of the load, it is necessary to assess parameter estimation during general movement.

Dynamic Estimation Using the PUMA

In the dynamic case, the joint position encoder and the wrist sensor data are sampled while the manipulator is in motion. A fifth order polynomial trajectory in joint space was used to minimize the mechanical vibrations at the beginning and the end of the movement, and to improve the signal to noise ratio in the acceleration data (Figure 4.3). For more popular bang-coast-bang type trajectories, the joint accelerations are zero except at the beginning and the end of the movements, resulting in poor signal to noise ratio in the acceleration data for most of the movement.

The PUMA robot lacked the acceleration capacity necessary to estimate the moments of inertia of the load. It also lacked velocity sensors at the joints, which made estimation of the acceleration of the load difficult. The dynamic estimates of mass and center of mass for the previous load are shown in the last column of Table 4.1. The data in these estimates were sampled while the manipulator was moving from $[0, 0, 0, -90, 0, 0]$ to $[90, -60, 90, 90, 90, 90]$ degrees on a straight line in joint space in 2 seconds. Joint 4 of the PUMA has a higher maximum acceleration than the other joints, and thus, a longer path was given for it. This movement was the fastest the PUMA can execute using the fifth order trajectory without reaching the maximum acceleration for any of its joints. The estimates used all 400 data points sampled during the 2 second movement. The results show a slight deterioration in these estimates when compared to the static estimates; but they are still within $40g$ and $2mm$ of the actual mass and displacement, respectively. The estimates for the moments of inertia are shown in the second column of Table 4.2. The signal to noise ratio in the acceleration and the force/torque data were too low for accurate estimates of the moments of inertia for this load ($0.00238 Kg \cdot m^2$ in the largest principal moment). In this case, the torque due to gravity is approximately 40 times greater than the torque due to the maximum angular acceleration of the load. Thus, even slight noise in the data would result in poor estimates of **I**.

Load Estimation

Parameters ($kg \cdot m^2$)	Actual Values	PUMA[1] Estimates	PUMA[2] Estimates	DDArm[1] Estimates
I_{11}	0.0244	0.0192	0.0246	0.0233
I_{12}	0	-0.0048	0.0006	0.0003
I_{13}	0	0.0019	0.0008	0.0007
I_{22}	0.0007	0.0021	0.0036	0.0001
I_{23}	0	-0.0016	-0.0004	-0.0002
I_{33}	0.0242	0.0176	0.0199	0.0236

[1] (all joints moving)
[2] (3 special test movements combined)

Table 4.2: Estimates of the moments of inertia.

Special Test Movements Using the PUMA

Experiments with a more substantial load were performed to improve estimates of the moments of inertia. The new experimental load is shown in Figure 4.2. This load has large masses at the two ends of the aluminum bar, resulting in large moments of inertia in two directions ($\sim 0.024 kg \cdot m^2$) and a small moment in the other. A typical set of estimates of the moments of inertia at the center of mass frame for the load with the gripper subtracted out is shown in Table 4.2 for the above all-joints-moving trajectory. They contain some errors but are fairly close to the actual values.

In order to improve the estimates, the data were sampled while the robot was following three separate 2-second rotational trajectories around the principal axes of the load. These trajectories used joint 4 and joint 6 only, and resulted in higher acceleration data than the previous trajectory, thus improving the signal to noise ratio in both the acceleration and the force/torque data. Typical estimates for these special movements show improvements over the estimates with the previous trajectory (Table 4.2). Although the estimate of I_{22} is slightly worse than before, all the other terms have improved; the cross terms, especially, are much smaller than before. Yet the **I** estimates are not as accurate as the estimates of the mass and the center of mass shown in Table 4.1. Most of the error is probably due to the large amount of noise present in the acceleration data caused by differentiating the joint angle data twice. Part of the error may be due to inaccuracies in the current kinematic parameters of the manipulator. Experiments have shown that the actual orientation of

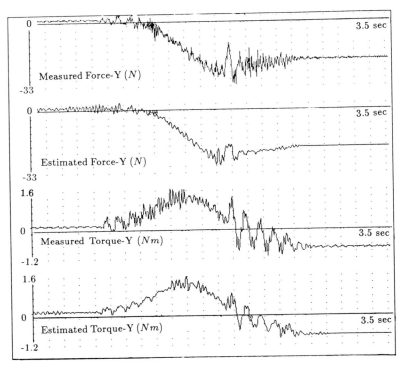

Figure 4.4: Measured force/torque data and computed force/torque data from the estimates using the PUMA.

the robot can be up to 4° off from the orientation computed from the encoder data.

Figure 4.4 shows the comparison of the measured forces and torques, and the computed forces and torques generated by a simulator from the estimated parameters and the measured joint data. The two sets of figures match very well even in the mechanical vibrations, verifying qualitatively the accuracy of the estimates. This suggests that for control purposes even poor estimation of moment of inertia parameters will allow good estimates of the total force and torque necessary to achieve a trajectory. This makes good sense in that the load forces with the PUMA are dominated by gravitational componenents, and angular accelerations experienced by the load are small relative to those components.

Load Estimation   79

4.2.2 MIT Serial Link Direct Drive Arm

The effect of the errors causing poor estimates of moment of inertia parameters could be alleviated by increasing the angular acceleration experienced by the load. Since the PUMA robot was already operating at its limits for the experimental results presented above, the algorithm was next implemented on the DDArm, which has a much higher acceleration and velocity capability. The DDArm also has a tachometer on each of its three joints so that numerical differentiation of positions is unnecessary; instead, the accelerations were obtained by digitally differentiating the velocities using a cutoff frequency of 30Hz. A Barry Wright Company Astek FS6-120A-200 6-axis force/torque sensor was used to measure all three forces and three torques about a point. The positions and velocities of the robot were initially sampled at 1kHz but were later down-sampled to match the sampling frequency of the force/torque sensor of 240 Hz. The identification procedure was again implemented off-line.

Dynamic Estimation Using the DDArm

The data used for estimating the inertial parameters of the load were sampled while the manipulator was moving from $(280°, 269.1°, -30°)$ to $(80°, 19.1°, 220°)$ in one second. Again a fifth order polynomial straight line trajectory in joint space was used. The resulting estimates for the moment of inertia parameters are shown in the last column of Table 4.2. The estimates for the mass and the location of the center of mass were as good as the PUMA results and are not shown. The estimated moment of inertia parameters are on the whole better than the PUMA results.

Parameters estimated for a set of special test movements using the direct drive arm were not substantially different and thus are not shown. The special test movements for the DDArm were not substantially faster than the movement of all joints, and thus probably contained the same amount of information.

Finally, Figure 4.5 shows the comparison of typical measured forces and torques with computed forces and torques generated by a simulator from the estimated parameters and the measured joint data. Once again there is a very good match between the measured and the predicted forces and torques. Thus, as expected, the combination of higher angular accelerations and good velocity sensing results in better parameter estimates.

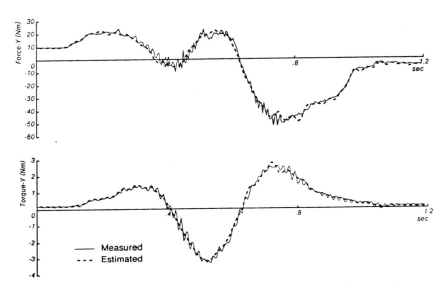

Figure 4.5: Measured force/torque data and computed force/torque data from the estimates using the DDArm.

4.3 Discussion

The work reported in this chapter has demonstrated that the inertial parameters of a manipulator load can be estimated accurately enough for purposes of control. The estimation algorithm was derived from the reformulation of the Newton-Euler equations for the rigid-body dynamics of a load so that the equations are linear in terms of the unknown inertial parameters. The estimation procedure then involved a simple least squares solution to a set of linear equations. In Chapter 5, this estimation method for a load will be extended to estimate the inertial parameters of all the links of a manipulator.

4.3.1 Usefulness of the Algorithm

It is important to realize that there are two distinct uses of an identified model. For control what matters is matching the input-output behavior of the model (in this case the relationship of load trajectory to load forces and torques) to reality, while for recognition/verification what matters is matching estimated parameters to a set of parameters postulated for reality. Both implementations of load inertial parameter estimation

Load Estimation 81

successfully match the input-output behavior of the load (Figures 4.4 and 4.5). The limited acceleration capacity of the PUMA robot and its limited sensing, however, result in relatively poor estimates of the moments of inertia of the load without the use of special test motions. In all cases the mass and the location of the center of mass could be accurately estimated from both series of static and dynamic measurements. Hence, identifying parameters well enough for recognition of the object may require large accelerations or special test movements in order to obtain the moment of inertia parameters accurately.

4.3.2 Sources of Error

This work is preliminary in that an adequate statistical characterization of the errors of the estimated parameters of the predicted forces has not been attempted. Nevertheless, some insights were gained into the sources of such errors.

The ultimate source of error is the random noise inherent in the sensing process itself. The noise levels on the position and velocity sensing are probably negligible, and could be further reduced using a model based filter such as the Extended Kalman Filter. Force and torque are measured by semiconductor strain gauges on structural members of the sensor. The random noise involved in such measurements is also probably negligible.

Bias Errors

Semiconductor strain gauges are notoriously prone to drift. Periodic recalibration of the offsets (very often) and the strain-to-force calibration matrix (often) may be necessary to reduce load parameter estimation errors further. During the experiments presented in this chapter, in order to minimize the bias errors, the data were taken after the force sensors had been left on for a while to warm up and the offsets were recalibrated before each data collection session.

Unmodeled Dynamics

A further source of noise is unmodeled structural dynamics. Neither the robot links nor the load itself are perfectly rigid bodies. A greater source of concern is the compliance of the force sensor itself. In order to generate structural strains large enough to be reliably measured with even semiconductor strain gauge technology, a good deal of compliance

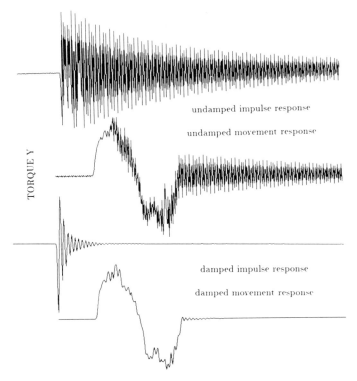

Figure 4.6: Vibration of load on force sensor.

is introduced into the force sensor. The load rigidly attached to the force sensor becomes a relatively undamped spring mass system. The response of the Astek force sensor to a tap on an attached load is shown in the "undamped impulse response" record of Figure 4.6. The effect of robot movement on this spring mass system is shown in the "undamped movement response" record.

There are several approaches that deal with this unmodeled dynamics. One approach is to attempt to identify the additional dynamics, which nevertheless greatly increases the complexity of the identification process and the amount of data that needs to be collected to get reliable estimates of any parameter.

Another approach is to try to avoid exciting the unmodeled dynamics by choosing robot trajectories that are as smooth as possible. The fifth

Load Estimation

order polynomial trajectories were chosen in the experiments so that the velocities and accelerations are always continuous. Using higher order polynomials would result in even greater smoothness. Nevertheless, with the PUMA a smooth commanded trajectory did not result in a smooth actual trajectory, because the control methods used and the actual hardware of the robot still introduced substantial vibration. One way to tell if the PUMA is turned on is to touch it and feel if it is vibrating. Vibrations were less of a problem with the direct drive arm, although still present.

The most successful approach is to mechanically damp out the vibrations by introducing some form of energy dissipation into the structure. Hard rubber washers were added between the force sensor and the load. The "damped impulse response" of Figure 4.6 illustrates the response of the force sensor to a tap on the load. The oscillations decay much faster with the added damping. The "damped movement response" indicates that this mechanical damping also greatly reduces the effect of movement on the resonant modes of the force sensor plus load.

A better method would have been to design appropriate damping into the force sensors, just as accelerometers are filled with oil to provide a critically damped response for a specified measurement bandwidth. Failing that, energy dissipation must be introduced either into the structural components of the robot or as a viscous skin on the gripper. As will be discussed later, appropriate mechanical damping may also be useful when using a force sensor in closed-loop force control.

Optimal Filtering

Finally, the need to differentiate the velocity numerically to find the acceleration greatly amplifies whatever noise is present. One can avoid the need to calculate acceleration explicitly by integrating equations (4.4) and (4.8). The derivations of the integrated estimation equations are included in Appendix 1. In our experiments with the integrated equations, the estimation results were not as good as with the original equations. Since an integrator is an infinite gain filter at zero frequency, large errors can result from small low frequency errors such as offsets. Therefore, the best performance will be achieved from applying some "optimal" filter, whose shape is probably an integrator at high frequencies but a differentiator at lower frequencies (Atkeson, 1986).

4.3.3 Inaccurate Estimates of the Moments of Inertia

One of the factors that make it difficult to identify moments of inertia accurately is the typically large contribution of the gravitational torque, which depends only on the mass and the relative location of the center of mass to the force sensing coordinate origin. A point mass rotated at a radius of $5cm$ from a horizontal axis must complete a full $360°$ rotation in 425 milliseconds for the torque due to angular acceleration to be equal to the gravitational torque, if a 5th order polynomial trajectory is used. A way to avoid gravitational torques is to rotate the load about the vertical axis, or to have the point of force/torque sensing close to the center of mass.

A simple example will illustrate the difficulty of recovering principal moments of inertia, given the moment of inertia tensor about the force sensing origin. The principal moment of inertia of a uniform sphere is only 2/7 of the total moment of inertia when it is rotated about an axis tangent to its surface, so that the effects of any errors in estimating the mass, the location of the center of mass, and the grip moments of inertia are amplified when the principal moment of inertia is calculated. This problem can be reduced by having the point of force sensing as close to the center of mass as possible

It still may be difficult to find the orientation of the principal moments of inertia even when the moment of inertia tensor about the center of mass has been estimated fairly accurately. This occurs when two or more principal moments of inertia are approximately equal. Finding the orientation of the principal axis is equivalent to diagonalizing a symmetric matrix, which becomes ill-conditioned when some of the eigenvalues are almost equal. A two dimensional example illustrates the problem. Consider the diagonalized matrix

$$\begin{bmatrix} \cos(\theta) & -\sin(\theta) \\ \sin(\theta) & \cos(\theta) \end{bmatrix} \begin{bmatrix} \lambda_1 & 0 \\ 0 & \lambda_2 \end{bmatrix} \begin{bmatrix} \cos(\theta) & \sin(\theta) \\ -\sin(\theta) & \cos(\theta) \end{bmatrix} \quad (4.21)$$

with eigenvalues $\{\lambda_1, \lambda_2\}$ and whose first principal axis is oriented at an angle θ with respect to the x axis. When the two eigenvalues are almost equal, the terms of the matrix dependent on the angle θ become very small. By substituting $\lambda_1 - \lambda_2 = \epsilon$ into the matrix (4.21),

$$\begin{bmatrix} \lambda_2 + \epsilon \cos^2(\theta) & \epsilon \cos(\theta)\sin(\theta) \\ \epsilon \cos(\theta)\sin(\theta) & \lambda_2 + \epsilon \sin^2(\theta) \end{bmatrix} \quad (4.22)$$

Load Estimation

All terms that contain angle information are multiplied by the difference (ϵ) of the principal moments of inertia. With a fixed amount of noise in each of the entries of the identified moment of inertia matrix, the orientation (θ) of a principal axis will become more and more difficult to recover as any pair of principal moments of inertia become approximately equal.

Chapter 5

Estimation of Link Inertial Parameters

The accurate values of the inertial parameters of the links of a manipulator are typically unknown even to the manufacturers of the manipulators. Determining these parameters from measurements or computer models is generally difficult and involves some approximations to handle the complex shapes of the arm components. This chapter presents a method of estimating all of the inertial parameters, the mass, the center of mass, and the moments of inertia, of each rigid body link of a robot manipulator using joint torque sensing.

The degree of uncertainty in inertial parameters is an important factor in judging the robustness of model-based control strategies. A common objection to the computed torque methods, which involve full dynamics computation (e.g., Luh, Walker, and Paul, 1980b), is their sensitivity to modeling errors, and a variety of alternative robust controllers have been suggested (Samson, 1983; Slotine, 1985; Spong, Thorp, and Kleinwaks; 1984, Gilbert and Ha, 1984). Typically these robust controllers express modeling errors as a differential inertia matrix and coriolis and gravity vectors without providing a reasonable basis for the source of errors or the bounds on errors. The error matrices and vectors combine kinematic and dynamic parameter errors, but as discussed in Chapter 3, kinematic calibration is sufficiently developed so that very little error can be expected in the kinematic parameters.

One aim of this work is to place similar bounds on inertial parameter errors by explicitly identifying the inertial parameters of each link that go into the making of the inertia matrix and coriolis and gravity vectors. The load identification results of Chapter 4 suggest, for example, that mass can be accurately identified to within 1%. Therefore, an assumption of 50% error in link mass in verifying a robust control formulation (Spong, Thorp, and Kleinwaks, 1984) is an unreasonable basis for argument. Slotine (1985) suggests that errors of only a few percent in inertial parameters make his robust controller superior to the computed torque method, but it may well be that these parameters can be identified more accurately than his assumptions.

In this book, as an alternative approach, the inertial parameters will be estimated on the basis of direct dynamic measurements. The algorithm of Chapter 4, used to identify load inertial parameters, can be modified to find link inertial parameters of a robot arm made up of rigid parts. The Newton-Euler dynamic equations are used to express the measured forces and torques at each joint in terms of the product of the measured movements of the rigid body links and the unknown link inertial parameters. These equations are linear in the inertial parameters. Unlike load estimation, the only sensing is one component of joint torque, measured either with strain gauges or from motor currents. This lack of full force/torque sensing, along with the restricted movement near the base, makes it impossible to find all the inertial parameters of the proximal links. As will be seen, these missing parameters have no effect on the control of the arm.

In this chapter, manipulators with only revolute joints are discussed, since handling prismatic joints requires only trivial modifications to the algorithm. The proposed algorithm was verified by implementation on the DDArm.

Previous Work

Mayeda, Osuka, and Kangawa (1984) required three sets of special test motions to estimate the coefficients of a closed-form Lagrangian dynamics formulation. The 10 inertial parameters of each link are lumped into these numerous coefficients, which are redundant and susceptible to numerical problems in estimation. On the other hand, every coefficient is identifiable since these coefficients represent the actual degrees of freedom of the robot. By sensing torque from only one joint at a time, their algorithm is more susceptible to noise from transmission of dynamic effects of distant links to the proximal measuring joints. For efficient dynamics computation, the recursive dynamics algorithms require the link parameters explicitly. Though recoverable from the Lagrangian coefficients, it is better to estimate the inertial parameters directly. Though this algorithm was implemented on a PUMA robot, it is difficult to interpret the results because of dominance of the dynamics by the rotor inertia and friction.

Mukerjee (1984; Mukerjee and Ballard, 1985) directly applied his load identification method to link identification, by requiring full force-torque sensing at each joint. Instrumenting each robot link with full force/torque sensing seems impractical, and is actually unnecessary given joint torque sensing about the rotation axis. Partially as a result, he does not address the issue of unidentifiability of some inertial parameters. This algorithm was later verified by simulation (Mukerjee, 1986).

Olsen and Bekey (1985, 1986) presented a two step link identification procedure using joint torque sensing based on the Newton-Euler equations. In their procedure for estimating the center of mass for each link, the angular accelerations were approximated to be 0's, restricting the movements to be slow for this phase. This restriction is not present in our method. Simulation results for two and three degree-of freedom robots were included, but no experiments were performed.

Khalil, Gautier, and Kleinfinger (1986) used a Lagrange formulation in presenting an identification model for link inertial parameters. They addressed the unidentifiability of some parameters, and used it to regroup the dynamic parameters and simplify computation. However, they showed no implementation results using their model.

Armstrong, Khatib, and Burdick (1986) measured the inertial properties of a PUMA 560 robot by counter-balancing the disassembled parts. This is an alternative approach to estimation, but is very tedious. Also the cross terms of the inertia matrix cannot be obtained in this way.

Khosla (1986; Khosla and Kanade, 1985) independently developed a link estimation algorithm very similar to ours. Working with the CMU DDArm II, he was also able to verify the algorithm by experiments.

Slotine and Li (1987) and Hsu et al. (1987) developed manipulator control algorithms which include on-line adaptation for rigid body link and load inertial parameters, without using the measurements of joint accelerations. Since the parameters of the links do not change once the manipulator is assembled, an off-line procedure as discussed in this chapter would be preferred in practice. Only the load parameters need to be estimated on-line.

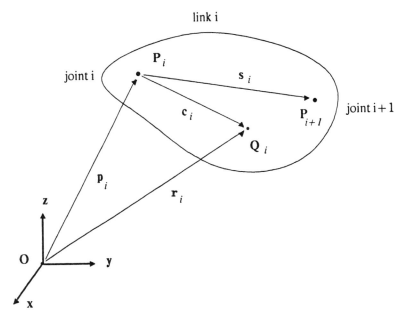

Figure 5.1: Coordinate origins and location vectors for link identification.

5.1 Estimation Procedure

5.1.1 Formulation of Newton-Euler Equations

In Chapter 4, the Newton-Euler equations for a rigid body load were formulated to be linear in the unknown inertial parameters, which were estimated by simple linear least squares. By treating each link of a manipulator as a load, this formulation can be extended to the link estimation problem. The differences in the equations are that only one component of force or torque is sensed and that the forces and torques from distal links are summed and transmitted to the proximal joints.

Consider a manipulator with n joints (Figure 5.1). Each link i has its own local coordinate system $\mathbf{P}_i$ fixed in the link with its origin at joint i. The joint force and torque due to the movement of its own link can be expressed by simply treating the link as a load and applying the equations from Chapter 4 for load identification:

Link Estimation

$$\begin{bmatrix} \mathbf{f}_{ii} \\ \mathbf{n}_{ii} \end{bmatrix} = \begin{bmatrix} \ddot{\mathbf{p}}_i - \mathbf{g} & [\dot{\omega}\times] + [\omega_i\times][\omega_i\times] & 0 \\ 0 & [(\mathbf{g} - \ddot{\mathbf{p}}_i)\times] & [\bullet\dot{\omega}] + [\omega_i\times][\bullet\omega_i] \end{bmatrix} \begin{bmatrix} m_i \\ m_i c_{x_i} \\ m_i c_{y_i} \\ m_i c_{z_i} \\ I_{xx_i} \\ I_{xy_i} \\ I_{xz_i} \\ I_{yy_i} \\ I_{yz_i} \\ I_{zz_i} \end{bmatrix}$$

or more compactly,

$$\mathbf{w}_{ii} = \mathbf{A}_i \boldsymbol{\phi}_i \tag{5.1}$$

where $\mathbf{w}_{ij}$ is the wrench (vector of forces and torques) at joint i due to movement of link j alone, $\mathbf{A}_i$ is the kinematic matrix that describes the motion of link i, and $\boldsymbol{\phi}_i$ is the vector of unknown link inertial parameters. All of the quantities are expressed in the local joint i coordinate system.

The total wrench $\mathbf{w}_i$ at joint i is the sum of the wrenches $\mathbf{w}_{ij}$ for all links j distal to joint i:

$$\mathbf{w}_i = \sum_{j=i}^{N} \mathbf{w}_{ij} \tag{5.2}$$

Each wrench $\mathbf{w}_{ij}$ at joint i is determined by transmitting the distal wrench $\mathbf{w}_{jj}$ across intermediate joints. This is a function of the geometry of the linkage only. The forces and torques at neighboring joints are related by

$$\begin{bmatrix} \mathbf{f}_{i,i+1} \\ \mathbf{n}_{i,i+1} \end{bmatrix} = \begin{bmatrix} \mathbf{R}_i & 0 \\ [\mathbf{s}_i\times]\cdot\mathbf{R}_i & \mathbf{R}_i \end{bmatrix} \begin{bmatrix} \mathbf{f}_{i+1,i+1} \\ \mathbf{n}_{i+1,i+1} \end{bmatrix} \tag{5.3}$$

or more compactly

$$\mathbf{w}_{i,i+1} = \mathbf{T}_i \, \mathbf{w}_{i+1,i+1} \tag{5.4}$$

where

$\mathbf{R}_i =$ the rotation matrix rotating the link $i+1$ coordinate system to the link i coordinate system,

$\mathbf{s}_i =$ a vector from the origin of the link i coordinate system to the link $i+1$ coordinate system, and

$\mathbf{T}_i =$ a wrench transmission matrix.

To obtain the forces and torques at the ith joint due to the movements of the jth link, these matrices can be cascaded:

$$\begin{aligned} \mathbf{w}_{ij} &= \mathbf{T}_i \mathbf{T}_{i+1} \cdots \mathbf{T}_j \mathbf{w}_{jj} \\ &= \mathbf{U}_{ij} \boldsymbol{\phi}_j \end{aligned} \quad (5.5)$$

where $\mathbf{U}_{ij} = \mathbf{T}_i \mathbf{T}_{i+1} \cdots \mathbf{T}_j \mathbf{A}_i$ and $\mathbf{U}_{ii} = \mathbf{A}_i$. A simple matrix expression for a serial kinematic chain (in this case a six joint arm) can be derived from (5.2) and (5.5):

$$\begin{bmatrix} \mathbf{w}_1 \\ \mathbf{w}_2 \\ \mathbf{w}_3 \\ \mathbf{w}_4 \\ \mathbf{w}_5 \\ \mathbf{w}_6 \end{bmatrix} = \begin{bmatrix} \mathbf{U}_{11} & \mathbf{U}_{12} & \mathbf{U}_{13} & \mathbf{U}_{14} & \mathbf{U}_{15} & \mathbf{U}_{16} \\ 0 & \mathbf{U}_{22} & \mathbf{U}_{23} & \mathbf{U}_{24} & \mathbf{U}_{25} & \mathbf{U}_{26} \\ 0 & 0 & \mathbf{U}_{33} & \mathbf{U}_{34} & \mathbf{U}_{35} & \mathbf{U}_{36} \\ 0 & 0 & 0 & \mathbf{U}_{44} & \mathbf{U}_{45} & \mathbf{U}_{46} \\ 0 & 0 & 0 & 0 & \mathbf{U}_{55} & \mathbf{U}_{56} \\ 0 & 0 & 0 & 0 & 0 & \mathbf{U}_{66} \end{bmatrix} \begin{bmatrix} \boldsymbol{\phi}_1 \\ \boldsymbol{\phi}_2 \\ \boldsymbol{\phi}_3 \\ \boldsymbol{\phi}_4 \\ \boldsymbol{\phi}_5 \\ \boldsymbol{\phi}_6 \end{bmatrix} \quad (5.6)$$

This equation is linear in the unknown parameters, but the left side is composed of a full force-torque vector at each joint. Since only the torque about the joint axis can usually be measured, each joint wrench must be projected onto the joint rotation axis (typically $[0, 0, 1]$ in internal coordinates), reducing (5.6) to

$$\boldsymbol{\tau} = \mathbf{K}\boldsymbol{\psi} \quad (5.7)$$

where $\tau_i = [0, 0, 0, 0, 0, 1] \cdot \mathbf{w}_i$ is the joint torque of the ith link, $\boldsymbol{\psi} = [\boldsymbol{\phi}_1, \boldsymbol{\phi}_2, \boldsymbol{\phi}_3, \boldsymbol{\phi}_4, \boldsymbol{\phi}_5, \boldsymbol{\phi}_6]^T$, and $\mathbf{K}_{ij} = [0, 0, 0, 0, 0, 1] \cdot \mathbf{U}_{ij}$ when the corresponding entry in (5.6) is nonzero. For an n-link manipulator, $\boldsymbol{\tau}$ is an $n \times 1$ vector, $\boldsymbol{\psi}$ is a $10n \times 1$ vector, and $\mathbf{K}$ is an $n \times 10n$ matrix.

5.1.2 Estimating the Link Parameters

Equation (5.7) represents the dynamics of the manipulator for one sample point. As with load identification, (5.7) is augmented using N data points:

$$\mathbf{K} = \begin{bmatrix} \mathbf{K}(1) \\ \vdots \\ \mathbf{K}(N) \end{bmatrix} \quad \boldsymbol{\tau} = \begin{bmatrix} \boldsymbol{\tau}(1) \\ \vdots \\ \boldsymbol{\tau}(N) \end{bmatrix}$$

Unfortunately, one cannot apply simple least squares estimation

$$\boldsymbol{\psi}_{estimate} = (\mathbf{K}^T \mathbf{K})^{-1} \mathbf{K}^T \boldsymbol{\tau} \quad (5.8)$$

Link Estimation

because $\mathbf{K}^T\mathbf{K}$ is not invertible due to loss of rank from restricted degrees of freedom at the proximal links and the lack of full force-torque sensing. Some inertial parameters are completely unidentifiable, while some others can only be identified in linear combinations.

Two different approaches were used to solve the above rank deficient problem. The simplest is ridge regression (Marquardt and Snee, 1975), which makes $\mathbf{K}^T\mathbf{K}$ invertible by adding a small number d to the diagonal elements:

$$\hat{\psi} = (\mathbf{K}^T\mathbf{K} + d\mathbf{I}_{10n})^{-1}\mathbf{K}^T\tau \qquad (5.9)$$

The estimates are nearly optimal if $d << \lambda_{min}(\mathbf{K}^T\mathbf{K})$, where λ_{min} is the smallest non-zero eigenvalue of $\mathbf{K}^T\mathbf{K}$.

Another approach expresses the dynamics in terms of a reduced set of inertial parameters that are independently identifiable and that allow the application of a straight least squares estimate. This reduced set can be generated either by examination of the closed form dynamic equations for linear combinations of parameters, or by application of singular value decomposition. Both methods were applied and the results checked against each other. The closed form dynamics equations were derived with the aid of MACSYMA (Mathlab Group, 1983) for the DDArm, since for 3 degrees of freedom the dynamic equations in closed form are already quite complicated. The results are summarized in Appendix 2 in terms of 15 essential variables; made explicit are both the unidentifiable parameters and the parameters identifiable only in linear combinations.

A far less complicated method that can be applied rather automatically to any manipulator kinematic structure is singular value decomposition of $\mathbf{K}$ in (5.8), yielding (Golub and Van Loan, 1983)

$$\mathbf{K} = \mathbf{U}\mathbf{\Sigma}\mathbf{V}^T$$

where $\mathbf{\Sigma} = \text{diag}\{\sigma_i\}$ and $\mathbf{U}$ and $\mathbf{V}^T$ are orthogonal matrices. For each column of $\mathbf{V}$ there corresponds a singular value σ_i which if not zero indicates that the linear combination of parameters, $\mathbf{v}_i^T\psi$, is identifiable. The unidentifiable parameters will have zero singular values associated with them. Since $\mathbf{K}$ is a function only of the geometry of the arm and the commanded movement, it can be generated exactly by simulation rather than by actually moving the real arm and recording data with the concomitant and inevitable noise. For completely unidentifiable parameters, the corresponding columns of $\mathbf{K}$ can be deleted without affecting τ. For parameters identifiable in linear combinations, all columns except one in

Parameters	Link 1	Link 2	Link 3
$m(Kg)$	67.13	53.01	19.67
$mc_x(Kg \cdot m)$	0.0	0.0	0.3108
mc_y	2.432	3.4081	0.0
mc_z	35.8257	16.6505	0.3268
$I_{xx}(Kg \cdot m^2)$	23.1568	7.9088	0.1825
I_{xy}	0.0	0.0	0.0
I_{xz}	-0.3145	0.0	-0.0166
I_{yy}	20.4472	7.6766	0.4560
I_{yz}	-1.2948	-1.5036	0.0
I_{zz}	0.7418	0.6807	0.3900

Table 5.1: CAD-modeled inertial parameters.

a linear combination can also be deleted. The resulting smaller $\mathbf{K}^T\mathbf{K}$ matrix will be invertible, and (5.8) can be used to estimate the reduced set of parameters.

5.2 Experimental Results

Link estimation was implemented on the DDArm. As discussed in Chapter 2, the ideal rigid body dynamics is a good model for this arm, uncomplicated by joint friction or backlash typical of other manipulators. Hence the fidelity of this manipulator's dynamic model suits estimation well. A set of inertial parameters is available for the arm (Table 5.1), determined by modeling with a CAD/CAM database (Lee, 1983). These values may not be accurate because of inevitable modeling errors, but they can serve as a point of comparison for the estimation results.

As discussed in Chapter 2, joint position is measured by a resolver and joint velocity by a tachometer. The tachometer output is of high quality and leads to good acceleration estimates after differentiation. The accuracy of the acceleration estimates plus high angular accelerations greatly improves inertia estimation. The joint torques are computed by measuring the currents of the 3 phase windings of each motor.

For the estimation results presented, 600 data points were sampled while the manipulator was executing 3 sets of fifth order polynomial trajectories in joint space. The specifications of the trajectories were:

Link Estimation

Parameters	Link 1	Link 2	Link 3
$m(Kg)$	0.0*	0.0*	1.8920†
$mc_x(Kg \cdot m)$	0.0*	-0.1591	0.4676
mc_y	0.0*	0.6776†	0.0315
mc_z	0.0*	0.0*	-1.0087†
$I_{xx}(Kg \cdot m^2)$	0.0*	4.1562†	1.5276†
I_{xy}	0.0*	0.3894	-0.0256
I_{xz}	0.0*	0.0118	0.0143
I_{yy}	0.0*	5.2129†	1.8967†
I_{yz}	0.0*	-0.6050†	-0.0160
I_{zz}	9.33598†	-0.8194†	0.3568

Table 5.2: Estimated inertial parameters.

1. $(330°, 289.1°, 230°)$ to $(80°, 39.1°, -10°)$ in 1.3 sec,

2. $(330°, 269.1°, -30°)$ to $(80°, 19.1°, 220°)$ in 1.3 sec,

3. $(80°, 269.1°, -30°)$ to $(330°, 19.1°, 220°)$ in 1.3 sec,

Since $\mathbf{K}^T\mathbf{K}$ in (5.9) is singular, estimates for the 30 unknowns are computed by adding a small number ($d = 10.0 << \lambda_{min}(\mathbf{K}^T\mathbf{K}) = 3395.0$) to the diagonal elements of $\mathbf{K}^T\mathbf{K}$.

Typical results under ridge regression are shown in Table 5.2. Parameters that cannot be identified because of constrained motion near the base are denoted by 0.0*. The first nine parameters of the first link are not identifiable because this link has only one degree of freedom about its **z** axis. These nine parameters do not matter at all for the movement of the manipulator and thus can be arbitrarily set to 0.0.

Other parameters marked by (†) can only be identified in linear combinations, indicated explicitly in Table 5.3. The ridge regression automatically resolves the linear combinations in a least squares sense. It can be seen that the estimated sums roughly match the corresponding sums inferred from the CAD-modeled parameters, but the sizeable discrepancy indicates that one parameter set may be more accurate than the other.

To verify the accuracy of the estimated and the modeled parameters, the measured joint torques are compared to the torques computed from the above two sets of parameters using the measured joint kinematic data. As shown in Figure 5.2, the estimated torques match the measured

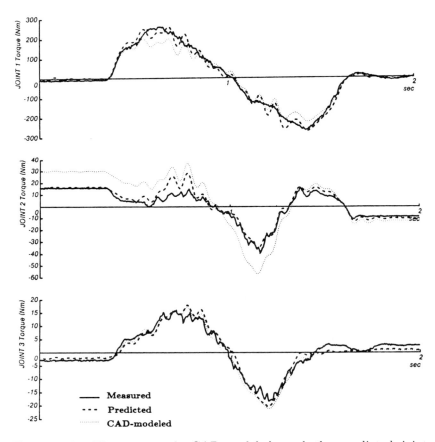

Figure 5.2: The measured, CAD-modeled, and the predicted joint torques.

Link Estimation

Linear Combinations	Estimated	CAD-Modeled
$m_3 c_{z_3} l_2 + I_{yz_2}$	-1.0710	-1.3526
$I_{xx_3} - I_{yy_3}$	-0.3691	-0.2702
$I_{zz_2} + I_{xx_3}$	0.7082	0.8632
$I_{zz_1} + I_{xx_2} + I_{xx_3} + m_3 l_2^2$	15.4236	13.0315
$I_{xx_2} + I_{xx_3} - I_{yy_2}$	0.4709	0.4147
$m_3 c_{z_3} - m_2 c_{y_2}$	-1.6863	-3.0814

Table 5.3: Parameters in linear combinations ($l_2 = 0.462\,m$.)

torques very closely. The torques computed from the CAD-modeled parameters do not match the measured torques as closely. This comparison verifies qualitatively that for control purposes the estimated parameters are in fact more accurate than the modeled parameters. Yet as seen in Chapter 4, one cannot make conclusion on the absolute accuracy of the estimates on the basis of the plots in Figure 5.2. The plots only tell us that one can predict the joint torques well.

Since the purpose of obtaining the estimates was to improve the control performance, the best test for verifying the quality of the estimates is to use them in a robot controller. In Chapter 6, the estimated inertial parameters will be used in studying the effectiveness of different control algorithms.

5.3 Identifiability of Inertial Parameters

There are three groups of inertial parameters: fully identifiable, identifiable in linear combinations, and completely unidentifiable. Membership of a parameter in a group depends on the manipulator's particular geometry. As shown in Table 5.2 and Appendix 2 for the DDArm, the 30 inertial parameters are grouped into the following categories:

1. fully identifiable: $m_2 c_{x_2}$, I_{xy_2}, I_{xz_2}, $m_3 c_{x_3}$, $m_3 c_{y_3}$, I_{xy_3}, I_{xz_3}, I_{yz_3}, I_{zz_3}

2. identifiable in linear combinations: I_{zz_1}, $m_2 c_{y_2}$, I_{xx_2}, I_{yy_2}, I_{yz_2}, I_{zz_2}, m_3, $m_3 c_{z_3}$, I_{xx_3}, I_{yy_3}

3. completely unidentifiable: m_1, $m_1 c_{x_1}$, $m_1 c_{y_1}$, $m_1 c_{z_1}$, I_{xx_1}, I_{xy_1}, I_{xz_1}, I_{yy_1}, I_{yz_1}, m_2, $m_2 c_{z_2}$

Some link inertial parameters are unidentifiable because of restricted motion near the base and the lack of full force-torque sensing at each joint. For the first link, rotation is only possible about its z axis. Suppose full force-torque sensing is available at joint 1. It can be seen from (5.1) that I_{xx_1}, I_{xy_1}, and I_{yy_1} are unidentifiable because they have no effect on joint torque. Since the gravity vector is parallel to the z axis, c_{z_1} is also unidentifiable. If it is now supposed that only torque about the z axis can be sensed, then all inertial parameters for link 1 become unidentifiable except I_{zz_1}.

In a multi-link robot a new phenomenon arises. Some parameters can only be identified in linear combinations, because proximal joints must provide the torque sensing to identify fully the parameters of each link. Certain parameters from distal links are carried down to proximal links until a link appears with a rotation axis oriented appropriately for completing the identification. In between, these parameters appear in linear combinations with other parameters. This partial identifiability and the difficulty of analysis become worse as the number of links are increased.

The ridge regression automatically resolves the linear combinations in a least squares sense, which makes these inertial parameters appear superficially different from those derived by CAD modeling. This is an approximation to computing the pseudoinverse to solve the rank deficient least squares problem.

Although not as simple as ridge regression, singular value decomposition of $\mathbf{K}$ in (5.8) to determine the minimal number of inertial parameters is attractive since it allows reformulating the dynamics with identifiable parameters only. The procedure isolates several sets of parameters whose linear combinations within each set are identifiable. The linear combinations can be reduced by consistently setting certain parameters in these sets to zero, leaving only one non-zero parameter in each set; for example, zeroing m_3, $m_3 c_{z_3}$, I_{xx_3}, and I_{xx_2} leaves I_{yz_2}, I_{yy_3}, I_{zz_2}, I_{zz_1}, I_{yy_2}, and $m_2 c_{y_2}$ as identifiable parameters. The unidentifiable parameters can also be set to zero. Finally, what remains is a reduced full-rank $\mathbf{K}^T \mathbf{K}$ matrix of dimension 15×15.

5.4 Discussion

Good estimates of the link inertial parameters were obtained as determined from the match of predicted torques to measured torques. The potential advantage of this movement-based estimation procedure for increased accuracy as well as convenience was demonstrated by the less accurately predicted torques based on the CAD-modeled inertial parameters.

The inaccuracy of the CAD-modeled parameters is due to several sources. The links and the motors are complicated, and computing the inertial parameters from the schematic drawing of the manipulator is bound to contain modeling errors. For the DDArm, the masses and moments of inertia are dominated by the large motors at the joints. The modeling of the inertial properties of these motors is difficult since the motors are made of complicated parts such as the stator windings. Also, the links can be attached to the rotor axes at arbitrary positions by the assembler, introducing uncertainty in the CAD-modeled parameters.

It is possible that the inaccuracy of the CAD-modeled parameters is exaggerated, since the same sensors that were used in the estimation are being used to compare the CAD-modeled parameters to the dynamically estimated parameters. Presumably a systematic error in the sensors, such as a mis-calibration of motor torque constants K_T, would be reflected in the dynamically estimated parameters. This would lead to a judgment of better match with these estimated parameters, even though the CAD-modeled parameters could conceivably be the more accurate. Ideally an independent measuring procedure such as weighing and counterbalancing should be used to resolve this point, but this was not tried.

With regard to errors in the motor torque constant, the motors were calibrated with a commercial force/torque sensor, and it is expected that errors in this calibrating device are very small. Problems of a dead zone near zero torque and torque ripple (Asada, Youcef-Toumi, and Lim, 1984) are not considered to be significant because of the large torques used in this study. Other sources of error are the same as discussed in the previous chapter, and are not repeated here.

Even supposing that there are possible errors in the sensors or kinematic variations due to assembly, the importance of the dynamic estimation of the link inertial parameters is actually emphasized. The controller must deal with the robot kinematics and sensor calibration as they exist, and to some extent the estimated model will accommodate kinematic variations and cancel sensor calibration error.

In our estimation experiments with the DDArm, fifth order trajectories were chosen so that joint accelerations are smooth and non-zero throughout the movements. The estimates of the inertial parameters could have been further improved if we had employed even more "exciting" trajectory commands. Recently, Armstrong (1987) developed a method of choosing optimal, persistently exciting trajectories for such identification experiments.

Chapter 6

Feedforward and Computed Torque Control

The accuracy of the manipulator dynamic model impinges on the performance of feedforward and computed torque control. Since friction is negligible for direct drive arms, and presuming that one has good control of joint torques, the issue of accuracy reduces to how well the inertial parameters of the rigid links are known. In Chapter 5, we developed an algorithm for estimating these inertial parameters for any multi-link robot as a result of movement and applied it to the DDArm. The accuracy of the inertial parameters was verified initially by comparing the measured joint torques to the torques computed from the estimated parameters. A more rigorous verification of the estimated model is in generating feedforward torques as part of a control algorithm. This chapter presents results of utilizing the estimated model to control the robot by both off-line (feedforward) and on-line (computed torque) computation of the joint torques.

Two sets of experiments were performed with the DDArm involving a subset of proposed control strategies. The first set of experiments is based on a hybrid control system. There is an independent analog servo for each joint with the position, velocity, and feedforward commands generated by a microprocessor. Since most commercial arms are controlled by a simple independent PID controller for each joint, an independent PD controller was tested on this arm to provide a baseline for comparison. The PD controller was augmented by feeding forward first gravity compensation and then the complete rigid body dynamics to ascertain any trajectory following improvements attained by taking the nonlinear dynamics more fully into account.

The second set of experiments shows the preliminary results of digital servo implementation, using one Motorola 68000 based microprocessor to control all the joints of the DDArm. The on-line computed torque approach is compared to the PD and to the feedforward approaches using the digital servo.

6.1 Control Algorithms

The full rigid body dynamics of an n degree-of-freedom manipulator are described by:

$$\tau = \mathbf{H}(\boldsymbol{\theta})\ddot{\boldsymbol{\theta}} + \dot{\boldsymbol{\theta}} \cdot \mathbf{C}(\boldsymbol{\theta}) \cdot \dot{\boldsymbol{\theta}} + \mathbf{g}(\boldsymbol{\theta}) \tag{6.1}$$

Since friction is small for the DDArm, it is ignored in the dynamics equation. The simplest and most common form of robot control is independent joint PD control, described by

$$\tau = \mathbf{K}_v(\dot{\boldsymbol{\theta}}_d - \dot{\boldsymbol{\theta}}) + \mathbf{K}_p(\boldsymbol{\theta}_d - \boldsymbol{\theta}) \tag{6.2}$$

where $\mathbf{K}_v$ and $\mathbf{K}_p$ are n×n diagonal matrices of velocity and position gains.

The feedforward controller augments the basic PD controller by providing a set of nominal torques τ_{ff}:

$$\tau_{ff}(\boldsymbol{\theta}_d, \dot{\boldsymbol{\theta}}_d, \ddot{\boldsymbol{\theta}}_d) = \hat{\mathbf{H}}(\boldsymbol{\theta}_d)\ddot{\boldsymbol{\theta}}_d + \dot{\boldsymbol{\theta}}_d \cdot \hat{\mathbf{C}}(\boldsymbol{\theta}_d) \cdot \dot{\boldsymbol{\theta}}_d + \hat{\mathbf{g}}(\boldsymbol{\theta}_d) \tag{6.3}$$

where the hat (ˆ) refers to the modeled values. When this equation is combined with (6.2), the feedforward controller results:

$$\tau = \tau_{ff}(\boldsymbol{\theta}_d, \dot{\boldsymbol{\theta}}_d, \ddot{\boldsymbol{\theta}}_d) + \mathbf{K}_v(\dot{\boldsymbol{\theta}}_d - \dot{\boldsymbol{\theta}}) + \mathbf{K}_p(\boldsymbol{\theta}_d - \boldsymbol{\theta}) \tag{6.4}$$

The feedforward term τ_{ff} can be thought of as a set of nominal torques which allow the dynamics (6.1) to be linearized about the operating points $\boldsymbol{\theta}_d, \dot{\boldsymbol{\theta}}_d$, and $\ddot{\boldsymbol{\theta}}_d$. Therefore, it is reasonable to add the linear feedback terms $\mathbf{K}_v(\dot{\boldsymbol{\theta}}_d - \dot{\boldsymbol{\theta}}) + \mathbf{K}_p(\boldsymbol{\theta}_d - \boldsymbol{\theta})$ as the control for the linearized system. These feedforward terms can be computed off-line, since they are functions of the parameters of the desired trajectory only.

On the other hand, the computed torque controller computes the dynamics on-line, using the sampled joint position and velocity data. The control equation is:

$$\tau_{ct}(\boldsymbol{\theta}_d, \boldsymbol{\theta}, \dot{\boldsymbol{\theta}}_d, \dot{\boldsymbol{\theta}}, \ddot{\boldsymbol{\theta}}_d) = \hat{\mathbf{H}}(\boldsymbol{\theta})\ddot{\boldsymbol{\theta}}^* + \dot{\boldsymbol{\theta}} \cdot \hat{\mathbf{C}}(\boldsymbol{\theta}) \cdot \dot{\boldsymbol{\theta}} + \hat{\mathbf{g}}(\boldsymbol{\theta}) \qquad (6.5)$$

where $\ddot{\boldsymbol{\theta}}^*$ is given by:

$$\ddot{\boldsymbol{\theta}}^* = \ddot{\boldsymbol{\theta}}_d + \mathbf{K}_v(\dot{\boldsymbol{\theta}}_d - \dot{\boldsymbol{\theta}}) + \mathbf{K}_p(\boldsymbol{\theta}_d - \boldsymbol{\theta}). \qquad (6.6)$$

If the robot model is exact, then each link of the robot is decoupled, and the trajectory error goes to zero. Gilbert and Ha (1984) have shown that the computed torque control method is robust to small modeling errors.

Previously, Liégeois, Fournier, and Aldon (1980) suggested the feedforward controller as an alternative to the on-line computation requirements of computed torque control, although they did not present any experimental results. Asada, Kanade, and Takeyama (1983) implemented a feedforward controller on the CMU Direct Drive Arm I, though for quite slow movements and for inertial parameters derived by CAD modeling.

The computed torque method has been considered by several other investigators (Bejczy, Tarn, and Chen, 1985; Luh, Walker, and Paul, 1980b; Markiewicz, 1973; Paul, 1972). Although simulation results have been presented, there have been very few published reports on the actual implementation of this controller, mainly due to the lack of an appropriate manipulator or on-line computational facility. In their computed-torque experiments with the PUMA 600, Leahy et al. (1986, 1987) have shown that for a highly geared manipulator, the forces not modeled by the Newton-Euler dynamics have a dominant effect in the trajectory tracking accuracy. Working with the CMU Direct Drive Arm II, Khosla and Kanade (1986) concluded that computed torque control is more accurate than independent-joint PD control. More recently, Khosla (1986) showed that the accuracy of feedforward controller was comparable to that of computed torque control.

6.2 Robot Controller Experiments

In this chapter, we first use the feedforward controller to evaluate the accuracy of our estimates of the link inertial parameters, and to compare its performance against several other simpler control methods for high speed movements. Then we present results on the implementation of a computed torque controller, again using the estimated inertial parameters of the links.

6.2.1 Analog/Digital Hybrid Controller

In this section, performances of several different controllers for full motion of the DDArm are evaluated using the hybrid digital/analog controller. The reference positions and velocities and the feedforward torques are generated by a microprocessor and are input to three independent analog joint servos. We present evaluations of the following five control methods used for high speed movements of all three joints of the manipulator:

1. PD controller with position reference only:

$$\boldsymbol{\tau} = -\mathbf{K}_v \dot{\boldsymbol{\theta}} + \mathbf{K}_p(\boldsymbol{\theta}_d - \boldsymbol{\theta})$$

2. PD controller with position reference and feedforward of gravity torques:

$$\boldsymbol{\tau} = \hat{\mathbf{g}}(\boldsymbol{\theta}_d) - \mathbf{K}_v \dot{\boldsymbol{\theta}} + \mathbf{K}_p(\boldsymbol{\theta}_d - \boldsymbol{\theta})$$

3. PD controller with position and velocity references:

$$\boldsymbol{\tau} = \mathbf{K}_v(\dot{\boldsymbol{\theta}}_d - \dot{\boldsymbol{\theta}}) + \mathbf{K}_p(\boldsymbol{\theta}_d - \boldsymbol{\theta})$$

4. PD controller with position and velocity references plus feedforward of gravity torques:

$$\boldsymbol{\tau} = \hat{\mathbf{g}}(\boldsymbol{\theta}_d) + \mathbf{K}_v(\dot{\boldsymbol{\theta}}_d - \dot{\boldsymbol{\theta}}) + \mathbf{K}_p(\boldsymbol{\theta}_d - \boldsymbol{\theta})$$

5. PD controller with position and velocity references plus feedforward of full dynamics:

$$\boldsymbol{\tau} = \hat{\mathbf{H}}(\boldsymbol{\theta}_d)\ddot{\boldsymbol{\theta}}_d + \dot{\boldsymbol{\theta}}_d \cdot \hat{\mathbf{C}}(\boldsymbol{\theta}_d) \cdot \dot{\boldsymbol{\theta}}_d + \hat{\mathbf{g}}(\boldsymbol{\theta}_d) + \mathbf{K}_v(\dot{\boldsymbol{\theta}}_d - \dot{\boldsymbol{\theta}}) + \mathbf{K}_p(\boldsymbol{\theta}_d - \boldsymbol{\theta})$$

The nominal position and velocity gains were adjusted experimentally to achieve high stiffness and overdamped characteristics without the feedforward terms.

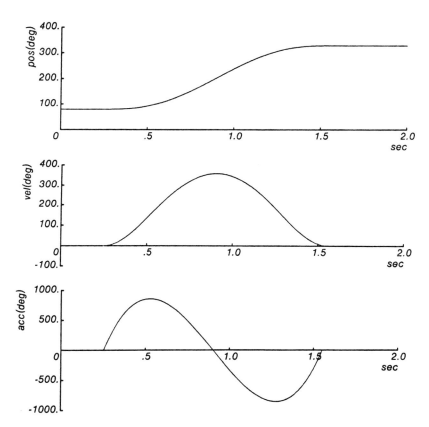

Figure 6.1: A fifth order polynomial trajectory.

A fifth order polynomial in joint space was used to generate the reference trajectory (Figure 6.1). The joints moved from $(80°, 269.1°, -30°)$ to $(330°, 19.1°, 220°)$ in $1.3s$, with peak velocities of 360 deg/sec and the peak accelerations of 854 deg/sec^2 for each joint. For control methods 2, 4, and 5, the estimates of the link inertial parameters given in Chapter 5 were used to compute the feedforward torques.

The trajectory errors for the above 5 controllers are shown in Figure 6.2. The errors for the first controller are very large and are out of the graph range. Adding a gravity feedforward term does not help very much, and the trajectory errors for Controller 2 are also very large. This was expected since gravity feedforward is a static correction to Con-

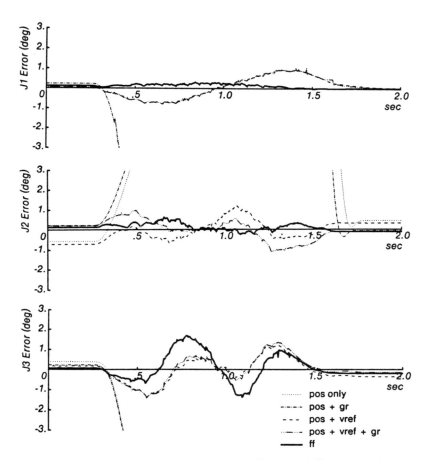

Figure 6.2: Trajectory errors of the 5 controllers for full 1.3s motion.

troller 1, and the dynamic effects dominate the response for high speed movements. Modifying the first controller by adding a velocity reference signal improved the response greatly. As with Controller 2, adding a gravity feedforward term did not reduce the trajectory errors very much, and influenced mainly the steady state errors for joints 2 and 3.

The full feedforward controller reduced the trajectory errors significantly for joints 1 and 2, with peak errors of only 0.33° and 0.64°, respectively. For joint 3, the feedforward torques did not help because of the light inertia and the dominance of unmodeled dynamics in the motor and in bearing friction. The high feedback gains make this joint somewhat robust to these unmodeled dynamics; yet, the trajectory errors could not be reduced below 1.4° with the feedforward torques based on the ideal rigid body model of the link.

6.2.2 Computed Torque Controller Experiment

In this section, results are presented for the computed torque method implemented on the DDArm. In this implementation, the analog servos are disabled, and the feedback computation is done digitally by one Motorola 68000 based microprocessor using scaled fixed-point arithmetic. Written in the C language, the controller, including the full computation of the robot dynamics, runs at a 133 Hz sampling frequency. Although further improvements in computation time are possible, this speed was adequate in demonstrating the efficacy of dynamic compensation. The details of this implementation are discussed in (Griffiths, 1986).

A similar fifth order polynomial trajectory as in the previous section was used for this experiment. Figure 6.3 shows the trajectory errors for three controllers: the digital PD controller, the feedforward controller using a digital servo, and the computed torque controller. The computed torque and the feedforward controllers both show a significant reduction in tracking errors for joints 1 and 2 compared with the PD controller, with no clear distinction between feedforward and computed torque. With the addition of dynamic components, the tracking errors are reduced from 4.4° to 2.2° for joint 1 and from 3.5° to 2.0° for joint 2. As before, the trajectory errors for joint 3 were not reduced by the computed torque or the feedforward controller. Again, this seems to indicate that our model for the third link may not be very good.

The trajectory tracking performance of the computed torque controller is not as good as that of the analog feedforward controller of the previous section. The main reason for this is the slow sampling frequency

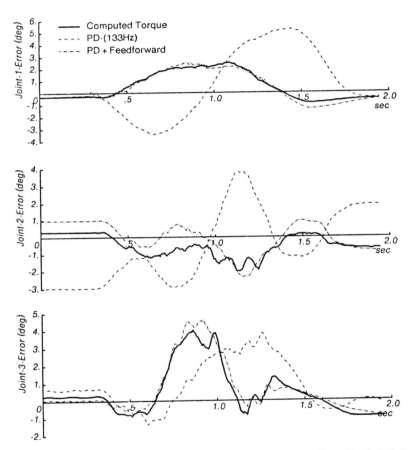

Figure 6.3: Trajectory errors of the three digital controllers for full 1.3s motion.

Feedforward and Computed Torque Control

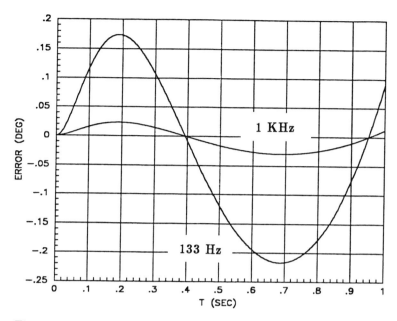

Figure 6.4: Trajectory errors for different sampling frequencies.

(133 Hz) of the digital controller, as compared to the 1 KHz sampling frequency at which the reference inputs were given to the analog servos. The effects of the two sampling frequencies on trajectory errors can be verified by simulation results for a simple one degree-of-freedom system with computed torque controller:

$$H\ddot{\theta} = \hat{H}(\ddot{\theta}_d + k_v(\dot{\theta}_d - \dot{\theta}) + k_p(\theta_d - \theta)) \tag{6.7}$$

For such a system, the computed torque controller is identical to the feedforward controller. The following parameters were used for the simulation:

$$H = \hat{H} = 10, \quad k_v = 20, \quad k_p = 100.$$

The trajectory errors (Figure 6.4) are shown for a 1 s duration fifth order polynomial trajectory travelling $250°$. The peak error for the controller with 133 Hz sampling frequency is shown to be approximately four times greater than with 1 KHz. This simulation result agrees with the experimental results in this section. Therefore, improvements in the computation time should also improve the tracking performance of the computed torque controller as well as the ability to use higher feedback gains (Khosla, 1987).

6.3 Discussion

We have presented experimental results of using an estimated dynamic model of the manipulator for dynamic compensation via feedforward and computed torque controllers. The results indicate that dynamic compensation by either model-based controller can improve trajectory accuracy significantly, when compared to independent joint PD control. Both feedforward and computed torque controllers performed similarly. For independent joint PD control, a velocity reference gave a substantially better result than not having a velocity reference. We did not, however, meet our goal of reducing trajectory following errors to the repeatability levels of the robot. There are several problems that became clear during this study: inconsistencies in the position and velocity sensors, torque estimation errors, and too low a sampling rate.

Inconsistencies in the position and velocity sensors: At high accelerations the position sensor, a resolver geared to the motor shaft, no longer is consistent with the velocity sensor, a tachometer mounted directly on the motor shaft. The integrated velocity signal is substantially different from the measured position signal. We believe that the resolver gearing leads to repeatable position sensing errors.

The consequence of position sensing errors are two-fold. At low trajectory error levels decreasing the position error will increase the velocity error and vice versa. The servo is unable to zero both errors simultaneously. Furthermore, position sensing errors cause commutation errors in the generation and measurement of torque in the electronically commutated brushless torque motors used in the robot. A remedy for this problem is an accurate position sensor built into the motor itself or directly attached to the motor shaft.

Torque estimation errors: The motors and the power amplifiers used in this robot do not accurately generate torques corresponding to the commanded torque levels. Measuring the currents in the three phases of the motor allows a more accurate estimate of torque to be made, but does not take into account nonlinearities in the current to torque relationship and errors in the commutation model. These types of torque errors are much more important than friction in this robot. Asada, Youcef-Toumi, and Lim (1984) addressed this problem and suggest incorporation of direct torque sensing in the structure of the motor.

Sampling rate: Increasing the sampling rate used in the feedforward and computed torque servos has two effects. The first effect is a reduction in digitization errors in generating the appropriate drive functions to the motors, and accurately modeling the rapidly changing nonlinear dynamics in the computed torque servo. The second effect is the increased level of feedback gains possible with a faster digital servo.

The results of the digital implementation of the feedforward and computed torque controllers were not as good as for the hybrid feedforward controller, because of a low digital sampling rate of 133 Hz. Recent improvements in our real-time computer architecture should allow the sampling rate to be increased several fold (Narasimhan et al., 1986). Nevertheless, we do not expect that an increased sampling rate will make the computed torque controller much more accurate than the feedforward controller. Khosla (1986) employed a sampling rate of 500 Hz, yet found that the feedforward controller and the computed torque controller were still equally accurate.

One goal of these experiments was to test the adequacy of the dynamically estimated rigid body model of the DDArm for control purposes. The results show that the model is quite accurate for joints 1 and 2, but not joint 3. The unmodeled dynamics of the light third link, including the motor dynamics and friction, are dominant and yield larger trajectory errors than at the other two joints. Better control of joint torque through improved motor modeling or direct torque sensing is probably the single most important requirement to improve trajectory following. We have found that the modeling errors of our rigid body model are extremely repeatable, and in Chapter 7 we present an adaptive feedforward control algorithm that compensates for modeling errors on repeated trajectories.

That feedforward and computed torque controllers were about equally accurate is not completely surprising, in that the same model is used in both controllers. The main difference in the two controllers is in their feedback controller mechanisms, and in the absence of large perturbations, the differences in the feedback controller may not be significant for trajectory following accuracy. If a robot is being used solely for free space movements without significant variation of its loads, then a hybrid controller using a fast independent analog servo in conjunction with digital feedforward terms may be quite adequate. In the case of large variation of loads or large disturbances, the pre-computed feedforward terms may not be accurate enough, and computed torque control with on-line adaptation for loads would give the best performance.

Chapter 7

Model-Based Robot Learning

An important component of human motor skill is the ability to improve performance by practicing a task. Commands are refined on the basis of performance errors. It is often suggested that such learning reduces the need for an accurate internal model, a model of the mechanical plant in the control system (e.g. Arimoto et al., 1984b; Wang and Horowitz, 1985; Harokopos, 1986a). This is not the case. Internal models play an important role in generating command corrections from performance errors. As an internal model is made more accurate, learning efficiency and initial performance are improved. This chapter will show how internal models can be used as learning operators using two examples: (1) positioning a limb at a visual target and (2) following a defined trajectory.

The type of learning described in this chapter – refining commands on the basis of practice – complements many other types of adaptive processes. Feedback controller designs can be improved by adaptive control algorithms. Internal models can be incrementally improved using system identification techniques. Trajectories can be optimized for particular tasks. Robot plans and programs can be debugged as errors are discovered during execution. This chapter focuses on improving execution of a given task plan by refining the commands given to the robot.

The model-based learning algorithms described here all have the same form. Commands are refined on the basis of performance errors. A command is applied to the controlled system (Figure 7.1A). Performance errors may result from errors in the command. A model of the inverse

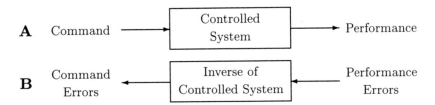

Figure 7.1: The inverse of the controlled system is used to estimate command errors from performance errors.

of the controlled system is used to estimate the errors in the command based on the measured performance or output errors (Figure 7.1B). If the inverse model of the controlled system is perfect, the command errors would be correctly estimated and completely eliminated after one attempt at performing the task. (Of course, if a perfect model of the controlled system is available, then the initial command would also have been perfect). Perfect knowledge of the controlled system is not usually available, and the model of the inverse of the controlled system will be incorrect. Due to the modeling errors, the command correction will be incomplete, and learning will be an iterative process of refining the command.

There are three steps to the learning algorithms: command initialization, execution, and modification. The initial command is generated by applying an inverse model of the controlled system to the desired performance. During execution, a command is applied to the system and the actual performance is monitored. The command correction is calculated by applying the inverse model to the performance errors. The refined command is now executed. The cycle of command execution and modification is repeated until desired performance is achieved.

In Section 7.1, kinematic learning is presented as an example of using models to correct performance errors. Trajectory learning is explored in the rest of the chapter. Section 7.2 introduces the trajectory learning problem, and Section 7.3 presents our model-based approach to perfecting performance on a single trajectory. Section 7.4 describes an implementation of the trajectory learning algorithm. Section 7.5 discusses the problems that may arise when using simplified models, and Section 7.6 explores convergence issues for trajectory learning. In our exploration of model-based learning we have used the kinematic and dynamic models described in Chapters 3, 4, and 5 of this book.

Model-Based Robot Learning 115

Figure 7.2: A robot arm and a target are viewed by a vision system.

7.1 Kinematic Learning

The task of positioning the limb at a visual target will be used to provide a specific example of how model-based learning works. A robot arm and a target are viewed by a vision system (Figure 7.2). The robot arm servos to a commanded set of joint angles $\boldsymbol{\theta}$ and the vision system measures the tip position $\mathbf{x}$ in vision system coordinates. The controlled system in this case transforms commanded joint angles into a measured tip position (Figure 7.1A):

$$\mathbf{x} = \mathbf{f}(\boldsymbol{\theta}) \qquad (7.1)$$

As mentioned in Chapter 1, the forward kinematics $\mathbf{f}(\boldsymbol{\theta})$ is in general a nonlinear transformation. For the purposes of this example we will assume that there are no singularities or redundancies to resolve in the field of view of the vision system. For each desired tip position there is one and only one appropriate set of joint angles.

A *model* of the inverse kinematics is used to transform the desired tip position, $\mathbf{x}_d$, into an initial joint angle command, $\boldsymbol{\theta}^0$, in the command initialization stage:

$$\boldsymbol{\theta}^0 = \hat{\mathbf{f}}^{-1}(\mathbf{x}_d) \qquad (7.2)$$

As in Chapter 1, a caret (ˆ) is used to indicate a model or an estimate of a quantity. The model may have been derived before the robot was built, or it may have been identified directly from measurements (Chapter 3).

The initial joint angle command is applied in the first execution stage, and the corresponding tip position is measured:

$$\mathbf{x}^0 = \mathbf{f}(\boldsymbol{\theta}^0) \tag{7.3}$$

The true system, $\mathbf{f}(\boldsymbol{\theta})$, and its inverse are unknown, and only imperfect models are available. Due to modeling errors, the actual tip position $\mathbf{x}^0$ will not match the desired tip position, $\mathbf{x}_d$.

At this point we must decide how to transform the measured tip position error into a correction to the set of commanded joint angles. Performance errors must be mapped into command corrections. The same model of the inverse kinematics that was used to generate the initial command, $\hat{\mathbf{f}}^{-1}()$, will be used to estimate the command error (Figure 7.1B).

The command error $\delta\boldsymbol{\theta}$ is the difference between the currently commanded joint angles $\boldsymbol{\theta}^0$ and the unknown correct set of joint angles, which will be indicated as $\boldsymbol{\theta}^*$. The command error can be computed in terms of the actual and desired performances using the true system inverse:

$$\delta\boldsymbol{\theta}^0 = \boldsymbol{\theta}^0 - \boldsymbol{\theta}^* = \mathbf{f}^{-1}(\mathbf{x}^0) - \mathbf{f}^{-1}(\mathbf{x}_d) \tag{7.4}$$

As we do not have perfect knowledge of the true system inverse, we must use a model of the system inverse to estimate the command error:

$$\widehat{\delta\boldsymbol{\theta}}^0 = \hat{\mathbf{f}}^{-1}(\mathbf{x}^0) - \hat{\mathbf{f}}^{-1}(\mathbf{x}_d) \tag{7.5}$$

The command is updated by simply subtracting the estimate of the command error from the previous command:

$$\boldsymbol{\theta}^1 = \boldsymbol{\theta}^0 - \widehat{\delta\boldsymbol{\theta}}^0 \tag{7.6}$$

If the model of the system inverse was perfect, the command error would be estimated correctly and completely eliminated on the next attempt. However, a model is rarely perfect, so command correction must be an iterative process of estimating a command error using an imperfect model, removing the estimated command error, applying the refined command, and using the resulting performance error and the model to estimate remaining errors in the command. Equations (7.3), (7.5), and (7.6) can be indexed with i to indicate that they are applied on each practice attempt, reflecting the iterative nature of the algorithm:

Model-Based Robot Learning

1. Command initialization:
$$\theta^0 = \hat{\mathbf{f}}^{-1}(\mathbf{x}_d) \tag{7.7}$$

2. Command execution:
$$\mathbf{x}^i = \mathbf{f}(\theta^i) \tag{7.8}$$

3. Command error estimation:
$$\widehat{\delta\theta}^i = \hat{\mathbf{f}}^{-1}(\mathbf{x}^i) - \hat{\mathbf{f}}^{-1}(\mathbf{x}_d) \tag{7.9}$$

4. Command modification:
$$\theta^{i+1} = \theta^i - \widehat{\delta\theta}^i \tag{7.10}$$

Steps 2, 3, and 4 are repeated until satisfactory performance is achieved.

Convergence for Kinematic Learning

The quality of the inverse model used as the learning operator determines how fast model-based learning converges. Fixed point theory can be used to analyze the general nonlinear case (Isaacson and Keller 1966, Wang 1984, Wang and Horowitz 1985). A learning algorithm can be viewed as a mapping of commands on the ith attempt to commands on the next attempt:

$$\theta^{i+1} = F(\theta^i) \tag{7.11}$$

The previously described algorithm can be put into this form by substituting equation (7.8) into (7.9) and (7.9) into (7.10). The model-based learning algorithm modifies the ith command by adding a correction based on the performance error transformed by the inverse model:

$$\theta^{i+1} = \theta^i - \left(\hat{\mathbf{f}}^{-1}(\mathbf{f}(\theta^i)) - \hat{\mathbf{f}}^{-1}(\mathbf{x}_d)\right) \tag{7.12}$$

Note that when the desired performance, $\mathbf{x}_d$, is achieved using the correct command, θ^*, then $\mathbf{f}(\theta^*) = \mathbf{x}_d$ and equation (7.12) reduces to the fixed point $\theta^{i+1} = \theta^i = \theta^*$.

We can ask whether this fixed point is stable by analyzing a linearization of equation (7.12) at the point $(\theta, \mathbf{x}) = (\theta^*, \mathbf{x}_d)$. For a small perturbation $\delta\theta$ from the fixed point,

$$\mathbf{f}(\theta^* + \delta\theta) = \mathbf{x}_d + \mathbf{J}(\theta^*)\delta\theta \tag{7.13}$$

where **J** is the Jacobian matrix, whose components are derivatives of **f**. Similarly, for a small perturbation $\delta\mathbf{x}$ from the fixed point,

$$\hat{\mathbf{f}}^{-1}(\mathbf{x}_d + \delta\mathbf{x}) = \hat{\mathbf{f}}^{-1}(\mathbf{x}_d) + \hat{\mathbf{J}}^{-1}(\mathbf{x}_d)\delta\mathbf{x} \qquad (7.14)$$

where $\hat{\mathbf{J}}^{-1}$ is the Jacobian matrix for the inverse model $\hat{\mathbf{f}}^{-1}()$. If on the ith trial the command is perturbed from $\boldsymbol{\theta}^*$ by $\delta\boldsymbol{\theta}^i$ so that $\boldsymbol{\theta}^i = \boldsymbol{\theta}^* + \delta\boldsymbol{\theta}^i$, the error in the next command, $\delta\boldsymbol{\theta}^{i+1} = \boldsymbol{\theta}^{i+1} - \boldsymbol{\theta}^*$, can be computed by substituting equations (7.13) and (7.14) into equation (7.12):

$$\delta\boldsymbol{\theta}^{i+1} = (\mathbf{1} - \hat{\mathbf{J}}^{-1}(\mathbf{x}_d)\mathbf{J}(\boldsymbol{\theta}^*))\delta\boldsymbol{\theta}^i \qquad (7.15)$$

In the linear case, the command error $\delta\boldsymbol{\theta}$ will decrease when all of the eigenvalues of the matrix $(\mathbf{1} - \hat{\mathbf{J}}^{-1}\mathbf{J})$ are less than one in absolute value, with the rate of decrease determined by the magnitude of the eigenvalues. If $\hat{\mathbf{J}}^{-1}$ is a correct inverse of **J** the command error will be completely corrected after one attempt. If the magnitude of any eigenvalue is greater than one, the learning process will be unstable and performance degraded rather than improved by learning. The magnitude of the eigenvalues of $(\mathbf{1} - \hat{\mathbf{J}}^{-1}\mathbf{J})$ depend on how accurately $\hat{\mathbf{J}}^{-1}$ inverts **J**, and thus the convergence rate of the learning algorithm depends on how closely the learning operator inverts the controlled system.

7.2 Trajectory Learning

In actual use robots tend to execute the same sequence of motions with the same loads repeatedly. We can take advantage of this pattern of usage by specializing the robot control system to store feedforward commands in a memory and play them back when necessary. This type of control system would repeat its errors on each movement, in contrast to the performance improvement with practice seen in human movement control. We propose an algorithm that uses practice to improve movement execution, by altering the stored feedforward commands on the basis of previous movement errors. This serves two purposes: 1) to improve robot trajectory following for repetitive movements, and 2) to increase our understanding of the role of practice in human motor control.

As discussed in Chapter 1, trajectory execution of a robot can be improved using a model-based learning algorithm. A model of the robot inverse dynamics is used as the learning operator that transforms trajectory following errors into feedforward command corrections. Model-based trajectory learning was implemented on the DDArm and greatly reduced trajectory following errors in a small number of practice movements.

Recent work in a number of laboratories has focused on how to refine feedforward commands for repetitive movements on the basis of previous movement errors (Arimoto 1985; Arimoto et. al. 1984abc, 1985; Casalino and Gambardella 1986; Craig 1984; Furuta and Yamakita 1986; Hara et al 1985; Harokopos 1986ab; Mita and Kato 1985; Morita 1986; Togai and Yamano 1985, 1986; Uchiyama 1978; Wang 1984; Wang and Horowitz 1985). These papers discuss only linear learning operators and emphasize the stability of the proposed algorithms. There has been little work emphasizing performance, i.e. the convergence rate of the algorithm. Section 7.6 describes the risks of only examining the convergence of a learning operator without assessing the transient performance. Poor transient performance during convergence usually implies a large sensitivity to disturbances and sensor and actuator noise.

What distinguishes this trajectory learning algorithm from previous robot trajectory learning schemes is the following combination:

- *Provides guidance for designing the learning operator:* The central problem in making use of trajectory error measurements is transforming those errors into feedforward command corrections. Previous work has explored the requirements on the learning operator for convergence, but has given little guidance as to how one should choose a particular learning operator from the large set of learning operators that converge. The algorithm proposed here explicitly uses an inverse plant model as the learning operator. If there are no disturbances or sensor or actuator noise and we have perfect models of the robot, the algorithm will correct any errors in the feedforward command after one practice motion.

- *Handles nonlinear plants:* The algorithm can make use of the nonlinear rigid body robot dynamics in the model used to generate feedforward command corrections. Previous work have used linearized or otherwise simplified robot models.

- *Makes explicit use of feedback control:* The algorithm takes explicit advantage of the on-line trajectory improvement provided by the feedback controller in calculating the next feedforward command.

- *Successful implementation:* The algorithm has been implemented on an actual robot, the MIT Serial Link Direct Drive Arm. Excellent trajectory learning performance requiring only a small number of practice trials has been achieved.

- *Generality (works with many feedback controllers):* The algorithm applies to a wide variety of feedback controller structures, as only knowledge of the feedback controller output is required. It would be easy to combine this adaptive feedforward control algorithm with adaptive feedback controllers.

- *Generality (works with many plants):* The algorithm does not require the plant dynamics to be of a particular type and applies to a wide range of robots with, for example, actuator dynamics, joint compliance dynamics, and flexible link dynamics. This is an important feature of the algorithm, as it is becoming clear that a typical robot exhibits quite complex dynamics (Good, Sweet and Strobel, 1985).

- *Analysis relevant to other schemes:* The analysis of this algorithm makes clear why other trajectory improvement schemes that intuitively seem correct often perform badly or actually degrade performance in practice.

7.3 The Trajectory Learning Algorithm

7.3.1 The Control Problem

Our goal is to execute repetitively an unrestrained robot trajectory with as small a trajectory following error as possible. The desired trajectory has a finite time duration and the robot starts in a known initial steady state $\mathbf{x}(0)$. We assume that modeling errors are much larger than any sensor and actuator noise or plant disturbances, and save the question of how to appropriately filter out non-repeatable noise and disturbances for further research.

There are three components of the trajectory learning algorithm: feedforward command initialization, movement execution, and feedforward command modification (Figure 7.3). After initializing the feedforward command, the movement execution and feedforward command modification steps are executed for each repeated movement attempt.

A: Feedforward Command Initialization

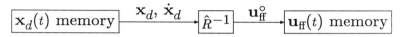

B: Movement Execution

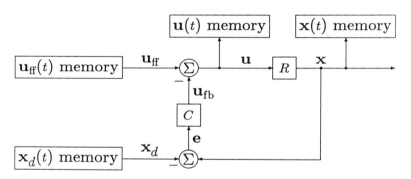

C: Feedforward Command Modification

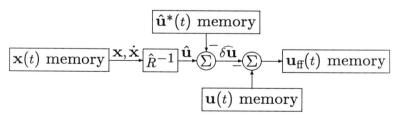

Figure 7.3: Block diagrams for feedforward command initialization, execution, and modification.

7.3.2 Feedforward Command Initialization

Given the desired state trajectory specification $\mathbf{x}_d(t)$ the feedforward command memory is initialized using a model of the inverse of the robot dynamics (Figure 7.3A):

$$\mathbf{u}_{\text{ff}}^0(t) = \hat{R}^{-1}(\mathbf{x}_d(t), \dot{\mathbf{x}}_d(t)) \tag{7.16}$$

The robot itself transforms a vector of actuator commands $\mathbf{u}$ into motion $\mathbf{x}$

$$\dot{\mathbf{x}} = R(\mathbf{x}, \mathbf{u}) \tag{7.17}$$

$\mathbf{x}$ represents the state vector of the robot, which includes the position and velocity of each joint ($\mathbf{x} = (\mathbf{q}, \dot{\mathbf{q}})$). A model of the inverse of the robot dynamics transforms a trajectory specification $\mathbf{x}(t)$ into actuator commands necessary to achieve the desired motion:

$$\mathbf{u} = \hat{R}^{-1}(\mathbf{x}, \dot{\mathbf{x}}) \tag{7.18}$$

For robot arms a rigid body dynamics model is often used to predict the forces and torques necessary to achieve a particular motion, and thus serves as a model of the inverted robot dynamics. For some robots it is argued that additional sources of dynamics are important (Goor, 1985a; Good, Sweet, and Strobel, 1985). In these cases we can still model the robot dynamics and invert the model. For the purposes of this chapter we will use only rigid body robot models, as these types of models describe most of the dynamics seen in a direct drive robot arm (Chapter 5).

7.3.3 Movement Execution

The ith attempt at the desired movement is executed using the current feedforward command $\mathbf{u}_{\text{ff}}^i(t)$ (Figure 7.3B). This command includes an error $\delta \mathbf{u}_{\text{ff}}^i(t)$ which is reduced by the actions of a suitably designed feedback controller C leaving an error $\delta \mathbf{u}^i(t)$ in the actuator commands sent to the robot $\mathbf{u}^i(t)$. For a single-input/single-output (SISO) linear plant and controller, the amount of error reduction can be expressed using Laplace transforms:

$$u^i(s) = \frac{1}{1 + C(s)R(s)} u_{\text{ff}}^i(s) \tag{7.19}$$

The actuator commands and the actually executed movement trajectory $\mathbf{x}^i(t)$ are stored for use by the learning module.

7.3.4 Feedforward Command Modification

The previous steps in this algorithm are standard components of many current robot controllers. One contribution of this work is to propose a particular approach to transforming trajectory errors into modifications of the feedforward command. We represent errors in our dynamic models of the robot as an input command disturbance, which we can correct for by modifying the feedforward command. We use our models of the inverted robot dynamics to estimate this actuator command error $\widehat{\delta \mathbf{u}}^i(t)$ which is then subtracted from the previously applied actuator commands to form the feedforward command for the next movement attempt (Figure 7.3C).

Estimating the actuator command error

We assume there exists a correct actuator command history $\mathbf{u}^*(t)$ that drives the plant exactly along the desired trajectory in the absence of disturbances or noise, and define the current actuator command error as

$$\delta \mathbf{u}^i(t) = \mathbf{u}^i(t) - \mathbf{u}^*(t) = R^{-1}(\mathbf{x}^i(t), \dot{\mathbf{x}}^i(t)) - R^{-1}(\mathbf{x}_d(t), \dot{\mathbf{x}}_d(t)) \quad (7.20)$$

where $R^{-1}()$ is the true inverse robot dynamics. Our estimate of the error in the current actuator command is found by simply replacing the true inverse robot dynamics with our model:

$$\begin{aligned}\widehat{\delta \mathbf{u}}^i(t) &= \hat{R}^{-1}(\mathbf{x}^i(t), \dot{\mathbf{x}}^i(t)) - \hat{R}^{-1}(\mathbf{x}_d(t), \dot{\mathbf{x}}_d(t)) \\ &= \hat{R}^{-1}(\mathbf{x}^i(t), \dot{\mathbf{x}}^i(t)) - \hat{\mathbf{u}}^*(t)\end{aligned} \quad (7.21)$$

If the plant is linear, we can apply the plant inverse directly to the trajectory error:

$$\hat{R}^{-1}(\mathbf{x}, \dot{\mathbf{x}}) - \hat{R}^{-1}(\mathbf{x}_d, \dot{\mathbf{x}}_d) = \hat{R}^{-1}(\mathbf{e}, \dot{\mathbf{e}}) \quad (7.22)$$

with $\mathbf{e} = \mathbf{x} - \mathbf{x}_d$. For the SISO linear case, the command error estimate can be expressed in terms of the feedforward command error using Laplace transforms.

$$e^i(s) = \frac{R(s)}{1 + C(s)R(s)} \delta u^i_{\text{ff}}(s) \quad (7.23)$$

$$\widehat{\delta u}^i(s) = \frac{1}{\hat{R}(s)} e^i(s) = \frac{1}{\hat{R}(s)} \frac{R(s)}{1 + C(s)R(s)} \delta u^i_{\text{ff}}(s) \quad (7.24)$$

Updating the feedforward command

The update for the feedforward command on the next movement is simply the modified actuator command:

$$\mathbf{u}_{\text{ff}}^{i+1}(t) = \mathbf{u}^i(t) - \widehat{\delta \mathbf{u}}^i(t) \tag{7.25}$$

By subtracting the correct command, $\mathbf{u}^*$, from both sides of Equation (7.25) and using Equations (7.19) and (7.24), we can express the Laplace transformed error propagation equation for the SISO linear case:

$$\delta u_{\text{ff}}^{i+1}(s) = \frac{\hat{R}(s) - R(s)}{\hat{R}(s)(1 + C(s)R(s))} \delta u_{\text{ff}}^i(s) \tag{7.26}$$

Although robot dynamics are typically multiple-input/multiple-output and nonlinear, an analysis of the single-input/single-output linear case gives us insight into how the algorithm behaves. With a perfect model, $\hat{R} = R$, the feedforward command errors will be zero after one movement. With an approximate model the feedforward command errors may only be reduced after one movement rather than eliminated. A sufficiently bad model may cause the feedforward command error to grow with practice, thus degrading performance rather than improving it.

7.4 Trajectory Learning Implementation

We have implemented the trajectory learning algorithm on the DDArm. To explore the effectiveness of our learning algorithm we will present results on learning a particular trajectory. All three joints of the DDArm were commanded to follow a fifth order polynomial trajectory with zero initial and final velocities and accelerations and a 1.5 second duration. Figure 7.4 shows the shape of the trajectory, which was the same for each joint, and Table 7.1 gives the initial and final joint positions, the peak joint velocities, and the peak joint accelerations.

An independent digital feedback controller was implemented for each joint and was not modified during learning. Since we are using a computer to generate the control signals, all signals and controlled systems were represented in discrete time rather than continuous time.

The initial feedforward torques were generated from a rigid body dynamics model. The model and the estimation of its parameters were described in Chapter 5. The calculated feedforward torques are shown in Figure 7.5A.

Model-Based Robot Learning

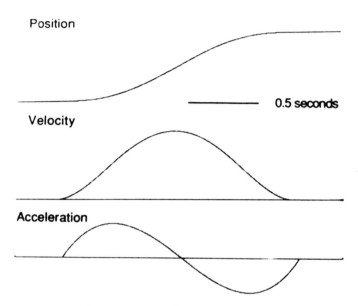

Figure 7.4: Test trajectory shape.

Joint	Initial Position radians	Final Position radians	Peak Velocity radians/s	Peak Acceleration radians/s^2
1	0.5	4.5	5.0	±10.3
2	5.0	1.0	-5.0	±10.3
3	4.0	-0.5	-5.6	±11.5

Table 7.1: Test trajectory parameters.

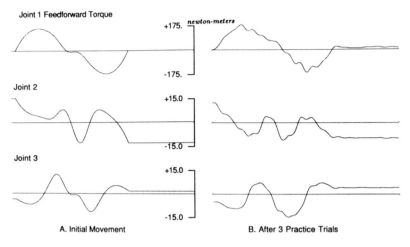

Figure 7.5: Feedforward Torques.

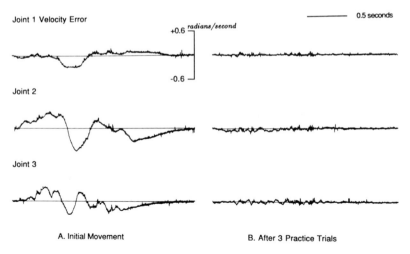

Figure 7.6: Velocity Errors.

As an index of trajectory following performance, the velocity errors (the difference between the actual joint velocity and the desired joint velocity) for the first movement are shown in Figure 7.6A. We have plotted the raw velocity error data to give an idea of the relative size of the trajectory errors and sensor noise.

In order to use the rigid body inverse dynamics model to compute joint torques, it was necessary to compute the joint accelerations. Joint positions and velocities were measured directly. A digital differentiating filter combined with an 8 Hz low pass filter was applied to the velocity data to estimate accelerations.

To reject noise and non-repeatable disturbances and to compensate for high frequency unmodeled dynamics, it was necessary to filter the trajectory errors and controller output. In this implementation we applied low pass digital filters with an 8 Hz cutoff to the data used in the learning process. We filtered the references used by the learning operator with the same filter used on the data. It was also necessary to correct for repeatable inconsistencies between the velocity sensors and the position measurements, which was done by adjusting the position reference to the feedback controller until the integrated velocity error matched the position error.

The robot executed two additional training movements which are not shown, and its performance on the fourth attempt of the test trajectory was assessed. Figure 7.5B shows the modified feedforward commands used on the fourth movement, and should be compared with the predicted torques shown in Figure 7.5A. Figure 7.6B shows the velocity errors for the fourth movement, and should be compared with the initial movement velocity errors in Figure 7.6A. There has been a substantial reduction in the trajectory following errors after only three practice movements.

7.5 Using Simplified Models

It may seem unnecessary to use the full rigid body dynamics model of the robot in the learning algorithm. One might think that learning algorithms of this type allow one to avoid modeling the robot in full detail. This is not the case. Simplifying the robot model necessarily introduces additional modeling error. Without careful analysis such modeling errors may cause the learning algorithm to have poor performance, or even to degrade performance. As an illustration of the possible effects of modeling error due to the use of simplified models, we will now present a seemingly reasonable simplified model of a two joint robot arm that, when used as the learning operator, fails to improve performance. The robot arm is a planar two link mechanism with rotary joints and is described in detail in (Brady, et. al., 1982, see Figure 9.2A).

The simplified dynamic model that we have chosen for this example is that of two independent rotary joints with constant moments of inertia. That is, we ignore the centripetal and Coriolis torques of the complete rigid body robot dynamics, and we assume constant moments of inertia around each joint. The moment of inertia of link 2 is a constant with respect to θ_2. The moment of inertia around joint 1 depends on θ_2. We approximate the moment around joint 1 as the average of the maximum and minimum moments around that joint, over all possible configurations of the robot. The equations of motion for such a simplified system are:

$$Torques = \hat{\mathbf{I}} \cdot \ddot{\boldsymbol{\theta}} \qquad (7.27)$$

where $\hat{\mathbf{I}}$ is a constant diagonal 2×2 matrix. This gives a learning operation of:

$$\mathbf{u}_{\text{ff}}^{i+1} = \mathbf{u}^i - \hat{\mathbf{I}} \cdot (\ddot{\boldsymbol{\theta}} - \ddot{\boldsymbol{\theta}}_d) \qquad (7.28)$$

The results of applying this learning algorithm to the simulated two joint robot movement are shown in Figure 7.7. The movement is from point **a** to point **b** with zero initial and final velocity, acceleration, and jerk (seventh order polynomial). Feedback control is provided by independent PD controllers at each joint, each having a bandwidth of 1.0 Hz and damping coefficient of 0.707 (based on $\hat{\mathbf{I}}$). Figure 7.7A shows the performance of the system on the initial trial, with the initial feedforward torques, $\mathbf{u}_{\text{ff}}^0$, based on the simplified dynamics model. Figure 7.7B shows the performance of the system after five iterations of the learning algorithm. In this case, where an inaccurate inverse model of the robot has been used to update the feedforward torque command, no improvement in trajectory following performance is seen. Had the full inverse dynamics model been utilized (under these ideal conditions of no measurement noise, actuator noise, or external torque disturbances), our algorithm would have produced a perfect movement after one iteration.

It has been argued that simplified models are appropriate for a robot with high gear ratios such as the PUMA. One must still model the other sources of dynamics prominent in these types of robots (Good, Sweet, and Strobel, 1985). Higher order actuator dynamics may play an important role (see, for example, Goor (1985a,b)). Our point is not that the rigid body dynamics are the only appropriate model and must be used, but

Model-Based Robot Learning

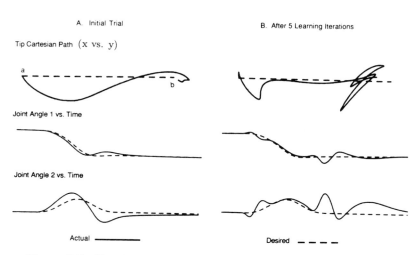

Figure 7.7: Simulated two joint learning with decoupled model.

that we must be careful to include all significant dynamics in our models. Learning performance can be used to assess the quality of the models used to drive the learning.

7.6 Trajectory Learning Convergence

An important question that arises is how close our model of the inverted robot dynamics has to be to the true inverted robot dynamics for the proposed trajectory learning algorithm to converge to zero trajectory error. We will indicate how convergence can be tested in the general nonlinear case. However, by specializing the convergence criteria to the linear case and presenting a numerical example, we will show that a convergence proof does not guarantee acceptable performance.

7.6.1 Nonlinear Convergence Criteria

One way to view the proposed trajectory learning algorithm is as a procedure to solve the vector nonlinear equations:

$$\chi_d = \mathcal{R}(\mu_{\text{ff}}) \qquad (7.29)$$

for μ_{ff}, where χ_d is the sampled version of the desired trajectory expressed as a vector $(\mathbf{x}_d[1], \mathbf{x}_d[2], \ldots, \mathbf{x}_d[T])^T$, μ_{ff} is a similar vector of the feedforward commands, T is the number of samples in the trajectory, and $\mathcal{R}$ is a nonlinear function representing the true robot and feedback controller dynamics. Since the initial state of the robot at the beginning of each movement is a known constant, $\mathcal{R}()$ does not require the robot initial state as an explicit argument.

One approach to analyzing convergence of algorithms to solve nonlinear equations is to combine the original equations and the proposed algorithm into a single iterative function. Each cycle of movement repetition and feedforward command modification can be expressed as an iterative function,

$$\mu_{\text{ff}}^{i+1} = \mathcal{G}(\mu_{\text{ff}}^i) \tag{7.30}$$

and we can appeal to fixed point theory for convergence conditions for μ_{ff} (Isaacson and Keller, 1966). This is also the approach suggested by (Wang, 1984; Wang and Horowitz, 1985).

7.6.2 Convergence Does Not Guarantee Good Performance.

Although conceptually elegant, the convergence conditions provided by fixed point theory for trajectory learning are too weak to be useful. The reason for this is that the duration of the executed trajectory is finite and the controlled plant is causal. The trajectory duration is referred to as the learning interval, as this is the time over which the feedforward command is modified. Convergence can be guaranteed by assuring that the feedforward command errors propagate forward in time on each cycle of movement execution and feedforward command modification until the command errors propagate out of the learning interval (Figure 7.8). Command and trajectory errors can grow exponentially as long as they are continually shifted forward in time, as we will show in the following example.

The example plant

Consider a rotary single degree of freedom robot. The dynamics of this robot in continuous time are

$$\tau = I\ddot{\theta} \tag{7.31}$$

Model-Based Robot Learning

where τ is the torque at the joint, I is the moment of inertia of the robot, and $\ddot{\theta}$ is the angular acceleration. We can express the dynamics of this robot in discrete time as

$$\theta_k - 2\theta_{k-1} + \theta_{k-2} = \frac{h^2}{2I}(\tau_k + \tau_{k-1}) \qquad (7.32)$$

where h is the sampling period (Åström and Wittenmark, 1984). Note that for later notational convenience in inverting the plant, the sampled torques have been renumbered to remove the plant delay. A sampling frequency of 1 kHz is used in the numerical simulations to follow.

We use a feedback controller

$$\tau_{fb_{k+1}} = -k(\theta_k - \theta_{d_k}) - b(\dot{\theta}_k - \dot{\theta}_{d_k}) \qquad (7.33)$$

to give the second order plant a natural frequency of 10 Hz and a damping ratio of 0.707. The angle and the angular velocity of the single joint are both measured directly.

Deriving the convergence criteria

Since the plant is linear, the inverse plant model used for feedforward command modification will be linear, and it can, therefore, operate directly on the trajectory error, $e[t]$. The example plant is also minimum phase and the plant inverse will be causal. In order to put convergence bounds on the learning operator we will refer to a general learning operator L instead of requiring the learning operator to be the inverse of the robot dynamics, $\hat{R}^{-1}$. L will be restricted to be a causal linear operator.

The propagation of feedforward command errors from one movement to the next is given by subtracting the correct command $u^*[t]$ from both sides of Equation (7.25):

$$\begin{aligned}\delta u_{\text{ff}}^{i+1}[t] &= \delta u^i[t] - \widehat{\delta u}^i[t] \\ &= \delta u^i[t] - l[t] * e[t]\end{aligned} \qquad (7.34)$$

where $l[t]$ is the impulse response of the learning operator L and $*$ is the convolution operator.

Since the learning interval (the duration over which we modify a particular feedforward command) is finite and includes only T samples, we can express Equation (7.34) as a matrix equation by replacing convolutions with matrix operations.

The sequence of sampled command errors and feedforward command errors are expressed as vectors, and the transformation between them can be expressed as a matrix equation $\delta \mathbf{u}^i = \mathbf{A} \delta \mathbf{u}_{ff}^i$:

$$\begin{pmatrix} \delta u^i[0] \\ \delta u^i[1] \\ \vdots \\ \delta u^i[T] \end{pmatrix} = \begin{pmatrix} a[0] & 0 & \cdots & 0 \\ a[1] & a[0] & \cdots & 0 \\ \vdots & \vdots & \ddots & \vdots \\ a[T] & a[T-1] & \cdots & a[0] \end{pmatrix} \begin{pmatrix} \delta u_{ff}^i[0] \\ \delta u_{ff}^i[1] \\ \vdots \\ \delta u_{ff}^i[T] \end{pmatrix} \quad (7.35)$$

where $\mathbf{A}$ is a lower triangular Toeplitz matrix. The elements $a[t]$ are the impulse response of the transfer function relating feedforward command errors to actuator command errors, $1/(1 + C[z]R[z])$, and we note that $a[0]$ must always be 1 due to a minimum loop delay of one sample before the feedback command can compensate for feedforward command errors.

The trajectory errors are generated by feedforward command errors according to the matrix equation $\mathbf{e}^i = \mathbf{B} \delta \mathbf{u}_{ff}^i$:

$$\begin{pmatrix} e^i[0] \\ e^i[1] \\ \vdots \\ e^i[T] \end{pmatrix} = \begin{pmatrix} b[0] & 0 & \cdots & 0 \\ b[1] & b[0] & \cdots & 0 \\ \vdots & \vdots & \ddots & \vdots \\ b[T] & b[T-1] & \cdots & b[0] \end{pmatrix} \begin{pmatrix} \delta u_{ff}^i[0] \\ \delta u_{ff}^i[1] \\ \vdots \\ \delta u_{ff}^i[T] \end{pmatrix} \quad (7.36)$$

where $b[t]$ is the impulse response corresponding to the transfer function $R[z]/(1 + C[z]R[z])$.

The effect of the learning operator can be expressed as a lower triangular Toeplitz matrix, $\mathbf{L}$, using the matrix equation $\widehat{\delta \mathbf{u}}^i = \mathbf{L} \mathbf{e}^i$:

$$\begin{pmatrix} \widehat{\delta u}^i[0] \\ \widehat{\delta u}^i[1] \\ \vdots \\ \widehat{\delta u}^i[T] \end{pmatrix} = \begin{pmatrix} l[0] & 0 & \cdots & 0 \\ l[1] & l[0] & \cdots & 0 \\ \vdots & \vdots & \ddots & \vdots \\ l[T] & l[T-1] & \cdots & l[0] \end{pmatrix} \begin{pmatrix} e^i[0] \\ e^i[1] \\ \vdots \\ e^i[T] \end{pmatrix} \quad (7.37)$$

Combining the previous matrix equations, Equation 7.34 can be written as the matrix equation:

$$\delta \mathbf{u}_{ff}^{i+1} = (\mathbf{A} - \mathbf{LB}) \delta \mathbf{u}_{ff}^i \quad (7.38)$$

Testing for convergence of the feedforward command error to zero reduces to guaranteeing that the absolute values of all the eigenvalues of the matrix $(\mathbf{A} - \mathbf{LB})$ are less than 1. Since $\mathbf{A}$, $\mathbf{L}$, and $\mathbf{B}$ are all lower

Model-Based Robot Learning

triangular Toeplitz matrices, the eigenvalues are all equal to the diagonal element of $(\mathbf{A} - \mathbf{LB})$, $a[0] - l[0]b[0]$. We recall that $a[0] = 1$, and note that the test for convergence in the finite learning interval case reduces to a test involving only the first element of the learning operator impulse response and the controlled plant impulse response:

$$|1 - l[0]b[0]| < 1 \tag{7.39}$$

This convergence test does not require L to be a very accurate inverse plant model.

A numerical example

To demonstrate that the above convergence condition does not guarantee acceptable performance let us examine a particular learning operator which satisfies the convergence test but generates bad performance. Let us choose a learning operator that satisfies the finite learning interval convergence test exactly such that

$$1 - l[0]b[0] = 0 < 1 \tag{7.40}$$

Such a learning operator is given by

$$l[t] = \begin{cases} \frac{1}{b[0]} & \text{if } t = 0 \\ 0 & \text{otherwise} \end{cases} \tag{7.41}$$

With this learning operator the eigenvalues of $(\mathbf{A} - \mathbf{LB})$ are all zero, and we are guaranteed exact convergence in at most T movements, where T is the number of samples in the learning interval. This learning operator corresponds to correcting the feedforward command with a scaled version of the position error history.

To examine the transient response of this learning operator we resort to numerical simulation. We initialize the feedforward command to have a single error on the first sample. The resulting position error is shown on the first graph of Figure 7.8. Several of the following movements are also shown, and by the 10th movement the position error has increased by a factor of 10^{26}. It is difficult to see that with each movement the first non-zero position error of the previous movement is exactly canceled, as the position errors later during the movement are blowing up exponentially. This exponentially growing wave of errors does shift forward in time, and after T movements will have shifted completely outside the

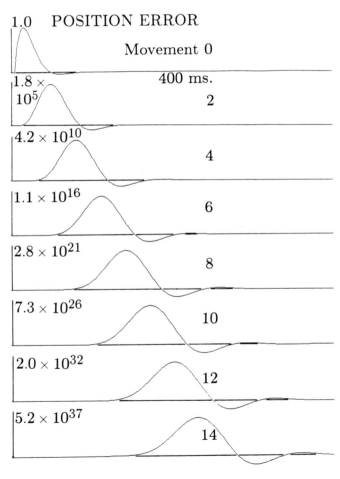

Figure 7.8: Simulated learning performance for a bad plant inverse model.

Model-Based Robot Learning

learning interval. The error within the T samples is zero after T practice movements. Had the correct inverse dynamics model been used as the learning operator (under these ideal conditions of no measurement noise, actuator noise, or disturbances) the feedforward command error would have been entirely canceled after one movement.

This type of finite time interval convergence relies on the absence of any introduction of new errors. In the presence of sensor and actuator noise and plant disturbances, the convergence promised by the finite time interval analysis will probably not occur. In addition, as the sampling interval length changes in the discretization of a continuous time plant, the convergence properties of this type of learning algorithm change. Convergence is improved when the plant is sampled at low frequencies, because the low sampling rate restricts the bandwidth of the learning operator.

A performance test

The previous example demonstrates the need to test performance in addition to showing convergence of a learning algorithm. In the nonlinear case performance must usually be checked by simulation and by actual implementation, but with linear plants and feedback controllers we can take advantage of superposition and the ability to transform to the frequency domain to check how errors at all frequencies are affected by the learning algorithm. This allows us to check learning convergence for an essentially infinite time interval, in contrast to the finite time convergence tests discussed previously (Equation 7.38).

From Equation 7.26 or 7.34 we see that the errors are propagated according to

$$\delta \mathbf{u}_{\text{ff}}^{i+1}[z] = \frac{1 - L[z]R[z]}{1 + C[z]R[z]} \delta \mathbf{u}_{\text{ff}}^{i}[z]$$
$$= \Lambda[z] \delta \mathbf{u}_{\text{ff}}^{i}[z] \qquad (7.42)$$

where the argument z indicates that we have shifted to the frequency domain using the Z transform. With a perfect inverse model, ($L = R^{-1}$), $\Lambda[z]$ is zero, implying convergence of the feedforward command error to zero after one practice movement. The learning algorithm will perform badly if, at any frequency, the gain of $\Lambda[z]$ is greater than 1. The gain of $\Lambda[z]$ as a function of frequency can be used to evaluate the efficacy of proposed learning operators. In the numerical example showing convergence with poor performance, the poor performance was indicated by a maximum gain of $\Lambda[z]$ equal to 530. This is approximately the factor by which the maximum position error was growing on each movement repetition in the simulation.

7.7 Discussion

The contributions of the work discussed in this chapter are:

- The derivation of a nonlinear learning algorithm based on explicit modeling of the controlled plant.

- The demonstration of good performance of the learning algorithm by implementation on the DDArm.

- The demonstration by a simple example that proofs of convergence do not guarantee acceptable performance and that is is necessary to verify performance.

The main message of this chapter is that models play an important role in learning from practice. Better models lead to faster correction of command errors. The incorporation of learning in a control system is not a license to do a poor modeling job of the controlled system. The benefits of accurate modeling are better performance in all aspects of control, while the risks of inadequate modeling are poor learning performance or even degradation of performance with practice.

The approach to robot learning presented here is based on explicit modeling of the robot and the task being performed. An inverse model of the task is used as the learning operator that processes the errors. Such model-based command refinement algorithms complement other approaches to adaptive control.

Some of the questions that warrant further research include the effect of modeling errors and non-repeatable disturbances on convergence, and learning of non-repetitive tasks. Reducing or filtering the estimated command correction will make model-based learning more robust to modeling errors. Convergence will be slowed, however. Further research is required into the appropriate tradeoff between handling modeling errors and fast convergence. Filtering of the model-based command update also plays an important role in reducing the effect of non-repeatable disturbances. If intermediate sensory signals are available, then breaking the control system into modules and having each module learn independently may improve learning performance. It is possible to modify models as well as commands during learning. In the examples presented in this chapter

Model-Based Robot Learning

the same models were used repeatedly even after it became clear during learning that the models had large errors.

The model-based learning algorithms are ideally suited to refining repetitive commands for the same tasks. The learning algorithms can also be applied to refining commands for different tasks by assuming that similar command errors will be made on similar tasks. An estimate of the command error on one task will be useful for improving the command for other tasks that share features with the original task.

Chapter 8

Dynamic Stability Issues in Force Control

To perform many useful tasks, robots must interact compliantly with the environment. The control of force may be demanded either as a response to uncertain contact conditions or as a requirement to complete a task. Uncertain contact conditions may arise from imprecision in modeling the environment or manipulator, or from tight tolerances in an assembly.

Though force control has long been recognized as an essential robot capability, actual implementations have not behaved satisfactorily on stiff environments (Whitney, 1987; Caine, 1985). Either the response is unstable, or it is slowed by high joint damping used to prevent instability. Stability is an essential property of any control system, yet many studies of robot force control have often neglected or treated stability as an afterthought.

Only during the last few years have there been a large number of papers discussing the stability problems associated with force control (Whitney, 1987; Roberts, 1984, 1985; Kazerooni, 1985, 1986a-c; Eppinger, 1986, 1987; Wlassich, 1986). Both Whitney (1987) and Roberts (1984) showed that a soft force sensor can lead to stable behavior with stiff environments. Drawbacks of a soft sensor include reducing the dynamic range of force response and the positional accuracy. Kazerooni (1985) presented a stability analysis and design method for an impedance controller using eigen-structure assignment. The complexity of his algorithm may preclude implementation for a multi-degree of freedom manipulator, and his experimental results have so far been on a single joint manipulator and a Cartesian (dynamically decoupled) manipulator.

Wlassich (1986) implemented an impedance controller on a two-link experimental manipulator, with the goal of making the manipulator behave as if it were a smaller mass (or inertia) than the actual mass. Wlassich discovered that unless the desired mass was larger than the actual mass, the manipulator became unstable against a stiff environment. Eppinger's analysis (1986) is similar to that contained in this chapter, but did not include a remedy for the stability problems.

The goal of this chapter is to present simple analyses to understand the dynamic stability problems associated with force control implementations, and then to present some design methods that would remedy those problems. Dynamic instability refers to instabilities caused by the interaction of the dynamics of the robot with the dynamics of the environment. Such instabilities can occur even for a single-link manipulator. Another type of instability is kinematic instability, caused by the kinematic coordinate transformations in the force control implementations, and is discussed in Chapter 9.

In Section 8.1, the stability problems are first identified by simple analysis and then verified by experiments. In the rest of the chapter, three different methods of improving the dynamic stability are presented. The method of using compliant coverings is discussed in Section 8.2, the adaptation to the environment stiffness in Section 8.3, and the use of joint torque control in Section 8.4. The stability analyses and the stable force control method are demonstrated by single-link experiments on the third link of the DDArm; multi-link cases are deferred until Chapter 9. As mentioned in Chapters 1 and 2, the direct drive arm is an ideal device for testing these ideas because the dynamics are close to ideal rigid body dynamics, since there is little friction and no backlash, and torques to the motors can be measured and controlled accurately.

8.1 Stability Problems

In discussing the dynamic stability problems of a force-controlled robot, two types of force control methods are used as examples: stiffness control (Salisbury, 1980) and impedance control (Hogan, 1985a-c). The stability problems, though, are general and not limited to these two methods.

Dynamic Stability Issues in Force Control

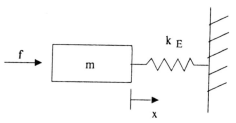

Figure 8.1: Model of the robot and the environment.

8.1.1 General Stability Analysis

To simplify the analysis, the robot is modeled as a mass m and the environment plus force sensor is modeled as a spring with stiffness k_E (Figure 8.1). Since the external force $f_{ext} = -k_E x$, the dynamic equation for this system is:

$$f - k_E x = m\ddot{x} \qquad (8.1)$$

Stiffness Control. In stiffness control (Salisbury, 1980), the interaction between the environment and the end-point of the robot is specified by a desired stiffness. For stiffness control of the above model, the control input is given by

$$\begin{aligned} f &= k_p(x_d - x) + k_v(\dot{x}_d - \dot{x}) + k_f(k_p(x_d - x) - f_{sensor}) & (8.2) \\ &= k_p(x_d - x) + k_v(\dot{x}_d - \dot{x}) + k_f(k_p(x_d - x) - k_E x) & (8.3) \end{aligned}$$

where k_p is the desired stiffness, k_v is the velocity gain, and k_f is the force feedback gain. Substituting (8.3) into (8.1), the dynamics of the total controlled system is described by

$$\begin{aligned} m\ddot{x} + k_v \dot{x} + (1 + k_f)(k_p + k_E)x &= k_p x_d + k_v \dot{x}_d + k_f k_p x_d & (8.4) \\ &= input\ terms \end{aligned}$$

Impedance Control. Impedance control (Hogan, 1985a-c) is a more general form of stiffness control. The interaction between the environment and the endpoint of the robot is specified by a desired mechanical impedance, which includes mass, stiffness, and damping terms. Impedance control, therefore, yields a form similar to (8.3), but with k_f depending on the desired apparent mass m_d. The control equation for the simple one degree of freedom robot is (Hogan, 1985b):

- Desired impedance:

$$m_d \ddot{x} + k_v(\dot{x} - \dot{x}_d) + k_p(x - x_d) = -f_{ext} \tag{8.5}$$

- Control input:

$$f = \frac{m}{m_d}(k_p(x_d - x) + k_v(\dot{x}_d - \dot{x})) + (\frac{m}{m_d} - 1)f_{ext} \tag{8.6}$$

$$= \frac{m}{m_d}(k_p(x_d - x) + k_v(\dot{x}_d - \dot{x})) - (\frac{m}{m_d} - 1)k_E x. \tag{8.7}$$

The force gain k_f is directly related to the desired mass m_d:

$$k_f = (\frac{m}{m_d} - 1) \tag{8.8}$$

If $m_d < m$, then k_f is positive and large. If $m_d = m$, the force sensor feedback is disabled. If $m_d > m$, then k_f is small but negative, resulting in positive force feedback. This, however, does not cause instability because of the $-k_E x$ term already present in the open loop system (8.1). The total system with the above impedance control is described by

$$m\ddot{x} + \frac{m}{m_d}k_v\dot{x} + (1 + k_f)(k_p + k_E)x = input\ terms \tag{8.9}$$

This equation is almost identical to (8.4) for the stiffness controlled system. In both cases, the closed loop equation includes a position term x which is multiplied by the effective stiffness of the environment k_E and the force feedback gain k_f. In fact, this form of equation will result from any force control method relying on a tip force sensor, since the output of the force sensor is essentially the stiffness k_E multiplied by the displacement x.

The system described by (8.4) or (8.9) is a stable system since all the poles have negative real parts given positive gains, k_v and k_p. In free space or in contact with a soft environment, k_E is small, and the manipulator will behave satisfactorily. For a stiff environment, $k_E \gg k_p$, and the system will be undesirably underdamped if k_v was computed with only

Dynamic Stability Issues in Force Control

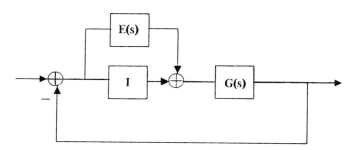

Figure 8.2: Error model for unmodeled dynamics.

k_p in mind, neglecting the large k_E term. Hence the combination of the force sensor and the stiff environment is approximately a very stiff spring, and from the stability point of view, force feedback is *very high gain* position feedback.

This linear and perfectly modeled system is underdamped but stable since the poles are still in the left half plane. Yet real systems are nonlinear, imperfectly modeled, and possibly unstable in the presence of unmodeled high frequency dynamics. Figure 8.2 shows a nominal model $G(s)$ of a force-controlled system and a multiplicative error model $E(s)$ for the unmodeled dynamics. $E(s)$ is not completely known and is only an approximation to the unmodeled dynamics. In a single input— single output system, for stability robustness with the above error model (Lehtomaki, 1981, see Appendix 3) shown in Figure 8.2, we need:

$$|E(s)| < |1 + G^{-1}(s)| \qquad (8.10)$$

For the combined manipulator-environment system, the loop transfer function is:

$$G(s) = \frac{sk_v + k_p(1 + k_f) + k_E k_f}{s^2 + k_E} \qquad (8.11)$$

Plots of $|E(s)|$ and $|1 + G^{-1}(s)|$ for different values of k_E are shown in Figure 8.3. The $|E(s)|$ plot is hypothetical, but typical of high-frequency unmodeled dynamics (Kazerooni, 1985).

For a value of k_v which gives an overdamped response in free space, $|I + G^{-1}(s)|$ has a large dip in the high frequency region for large values of k_E. Since the unmodeled dynamics are greater for higher frequencies, a non-robust system results. From the figure, $|E(s)| > |I + G^{-1}(s)|$ for

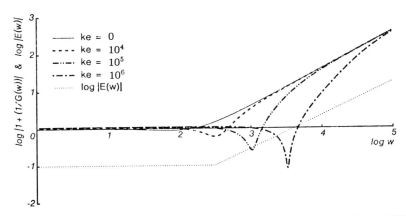

Figure 8.3: $|I + G^{-1}(s)|$ for $k_v = 250$, $m = 1$, $k_f = 10$, and $k_p = 1000$. Typical $|E(s)|$.

$k_E > 10^5$ Nt/m, and environments with this or greater stiffness may lead to unstable contact. For a soft environment, the figure shows that the system remains stable in the presence of unmodeled dynamics. This is precisely the behavior observed by previous investigators, as mentioned in the introduction.

In summary, three problems affect stability for a robot in contact with the environment:

1. force sensor feedback is essentially high gain position feedback,

2. there are always unmodeled high frequency dynamics in the robot system, and

3. the robot must deal with stiff environments.

8.1.2 Example of Unmodeled Dynamics

The general analysis of the previous subsection is verified in this section with a slightly more complex model of a robot. Even single degree-of-freedom robots are not completely rigid and possess flexible modes. Adding a flexible mode to the system of Figure 8.1 yields a fourth order model of the robot in contact with the environment as in Figure 8.4.

If a force controller were designed straightforwardly as a single mass system neglecting the flexibility, the root locus of the closed-loop system

Dynamic Stability Issues in Force Control

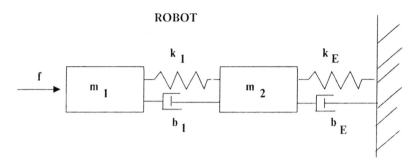

Figure 8.4: Two-mass model (flexibility).

in contact with a stiff environment ($k_E = 8 \times 10^5 \ N/m$) would behave as shown in Figure 8.5A. As the force gain k_f is increased, the poles move to the right half plane and the system becomes unstable. When the robot is in contact with a soft environment ($k_E = 10^3 \ N/m$), the closed-loop poles do not move to the right half plane for the same force gains. These two cases agree with the behavior predicted by the general analysis of the previous section.

8.1.3 Experimental Verification of Instability

Experimental results on the third link of the DDArm verify the previous analyses. The third link is controlled by pure force control, as shown on the block diagram of Figure 8.6. A Barry Wright FS6-120A 6-axis force/torque sensor was used to measure the contact force. As the block diagram shows, the tip force sensor feedback is a pure gain and has no dynamic compensation. No attempt is made here to reduce the instabilities that are present, although such an attempt is made later in this chapter.

Figure 8.7 shows the step responses of the simple force controller on three different surfaces using the same gains and inputs. The negative bias shown in the top plot is from an offset drift in the force sensor and should be ignored. As expected, the robot becomes unstable when it comes in contact with a stiff aluminum surface. The spikes in the force data are produced as the robot bounces on this surface. Stability is improved during contact with a rubber surface, but since this rubber happened to be hard, the response is still quite underdamped. The last plot of Figure 8.7 is the force step response of the robot in contact with

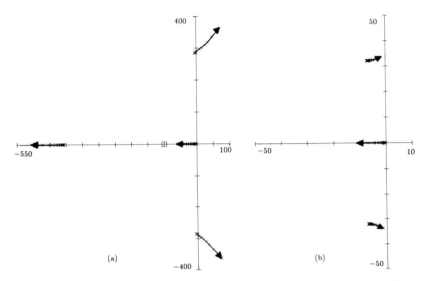

Figure 8.5: Root locus for two-mass model as k_f varies for (a) stiff and (b) soft environment.

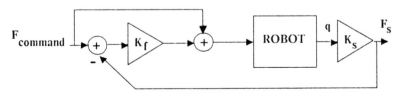

Figure 8.6: Simple force controller with $K_f = 0.222$.

Dynamic Stability Issues in Force Control 147

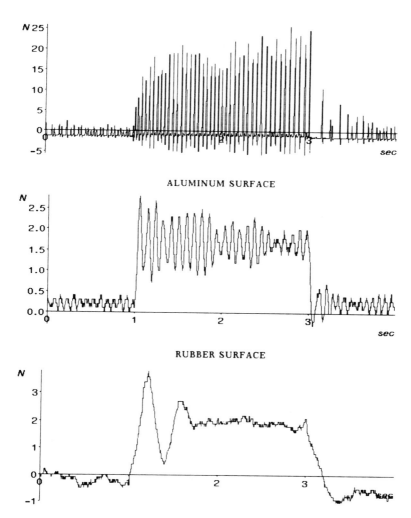

Figure 8.7: Step responses on three surfaces.

one author's fingers. Since human fingers and hands are very compliant, the robot is stable under simple force control.

This section has shown that a robot under simple force control may be unstable against a stiff environment. In the next three sections, several methods of remedying such stability problems are presented.

8.2 Compliant Coverings

As shown in equations (8.4) and (8.9), the force sensor feedback is essentially a position feedback with the gain dominated by the stiffness k_E, representing the effective stiffness of the force sensor and the environment. In general, the stiffness k_s of a wrist force sensor is very high so that it has a good force resolution and dynamic range. Since the sensor mass is typically small, the effective stiffness k_E is a result of serial combination of the sensor stiffness k_s and the environment stiffness k_{env}. Then, to reduce k_E, one can either make the sensor soft or make the environment appear soft by attaching a compliant covering to the contact surface. Either approach gives the same result and the root locus of such a system will have the pattern shown in Figure 8.5B. The experimental result for such a system can be seen in Figure 8.7 in the force step response on rubber or finger surface.

Whitney (1987) and Roberts (1985) have both suggested using soft sensors, but that may not be very practical since the softer the sensor is, the more the sensor will bend. Then the dynamic range of the sensor will be limited and it would also be difficult to control the tip position accurately due to large sensor deflections (Roberts, 1985). On the other hand, if a thin compliant covering is used on top of the force sensor, the environment will always seem soft to the force controller. This method preserves the large dynamic range of a stiff force sensor and also improves stability.

Using compliant coverings is not unlike how human arms are structured. In fact, the skin on the human arm helps us greatly in force control. It absorbs impact and it also acts as a contact sensor to give relative position information. In that respect, it would be even better to make the compliant coverings serve as tactile sensors, such as the types discussed in (Siegel, Garabieta, and Hollerbach, 1986).

Dynamic Stability Issues in Force Control

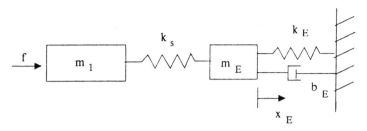

Figure 8.8: Fourth order model of the robot and the environment.

8.3 Adaptation to the Environment Stiffness

Another method in dealing with a stiff environment is to estimate the environment stiffness and then to incorporate the knowledge of the environment in the controller to achieve a stable and robust behavior. One simple method of incorporating such knowledge is to vary the velocity gains according to the environment stiffness and damping, so that the system remains stable without being too sluggish. The problem, then, is first to identify the dynamic characteristics of the environment.

8.3.1 Modeling

Consider a fourth order model for a situation where a manipulator is in contact with its environment (Figure 8.8). The force sensor is modeled as a spring of stiffness k_s. The environment is modeled as a spring, a damper, and a mass. The mass m_E is the manipulator load, such as a tool. In actual situations, if the system becomes unstable, the manipulator will lose contact with the environment. In this chapter, however, we consider only the simplified situation where the manipulator remains in contact.

Ignoring the viscous damping in the sensor, the equation for the dynamics of the environment is,

$$f_s = k_s x_s = m_E \ddot{x}_E + b_E \dot{x}_E + k_E x_E \qquad (8.12)$$

In this simple model, three parameters, m_E, b_E, and k_E, describe the dynamics of the interacting environment. For stability purposes, the stiffness parameter k_E is the most important among the three.

The measurement of f_s is available using a wrist force/torque sensor. Although x_E cannot be measured directly, it is related by

$$x_{robot} - x_s = x_E \qquad (8.13)$$

where x_{robot} is the movement of the manipulator from the point of contact with the environment and $x_s = f_s/k_s$. Typically we will not be able to measure the other state, $\dot{x}_E$.

8.3.2 Least Squares Algorithm

If the measurements of $\ddot{x}_E$ and $\dot{x}_E$ as well as f_s and x_E were available, then the above three parameters can be estimated straightforwardly by least squares on the continuous time equation (8.12); but at best, there are only measurements of f_s and x_E. By a number of different methods (Franklin and Powell, 1980), the continuous time equation can be transformed to a discrete version whose coefficients can be estimated by least squares. The actual continuous time parameters can then be obtained from these discrete coefficients by simple algebraic transformations. In the rest of this section, these steps are derived, and both simulation and experimental results are presented.

Derivations

The Laplace transform of the continuous time equation (8.12) for the dynamics of the environment is

$$f_s(s) = k_s x_s(s) = (m_E s^2 + b_E s + k_E)x_E(s). \qquad (8.14)$$

For the bilinear transformation method of converting a continuous time system to the discrete time domain, the relation between s and z is given by

$$s = \frac{2}{T}\left(\frac{z-1}{z+1}\right). \qquad (8.15)$$

Substituting (8.15) into (8.14),

$$\frac{x_E(s)}{f_s(s)} = \frac{1 + 2z^{-1} + z^{-2}}{a_0 + a_1 z^{-1} + a_2 z^{-2}} \qquad (8.16)$$

The difference equation corresponding to (8.16) is

$$f_s[n] + 2f_s[n-1] + f_s[n-2] = a_0 x_s[n] + a_1 x_s[n-1] + a_2 x_s[n-2]. \qquad (8.17)$$

Dynamic Stability Issues in Force Control

The coefficients a_0, a_1, and a_2 can be estimated by least squares, and the continuous time parameters can then be computed by:

$$b_E = \frac{T}{4}(a_0 - a_2)$$

$$m_E = \frac{T^2}{16}(a_0 + a_2 - a_1)$$

$$k_E = \frac{1}{2}(a_1 + \frac{8m_E}{T^2}) \tag{8.18}$$

Since the bilinear transformation is not an exact transformation, this derivation is approximate but quite accurate for a fast sampling time.

Simulation

The simulation results of the recursive least squares estimation is shown in Figure 8.9. The signal is a step response with some random noise and the sampling frequency is $200\,Hz$. As shown in the figure, although the stiffness estimate reaches the actual value quickly, within 0.05 second or 10 samples, the estimate for the mass does not settle to the final value until after 0.2 seconds. Also, because of the approximations in the discretization, there is significant bias error in this estimate. These behaviors are expected. The b_E and m_E parameters are mainly affected by the $\dot{x}_E$ and $\ddot{x}_E$ signals respectively, whose measurements are not available directly. Therefore, the discrete algorithm is essentially numerically differentiating and double differentiating the x_E measurements in order to arrive at estimates of b_E and m_E. The estimates for these parameters then should not be as good as the estimate of k_E.

Experiments

Least squares estimation was tested on two different surfaces, aluminum and rubber, using the third link of the DDArm. Figures 8.10 and 8.11 show the step force responses, the position displacements, and the estimation results. A resolver was used in measuring joint angle displacement. Since the stiffness of the force sensor is very high ($>> 10^6 N/m$), deflections in the sensor were neglected.

The results show that the estimated value of stiffness for the hard rubber surface is 60% of that for the aluminum surface. Although the actual stiffness of aluminum may be much higher than the estimate, the effective stiffness of the aluminum surface and the end effector (a bearing

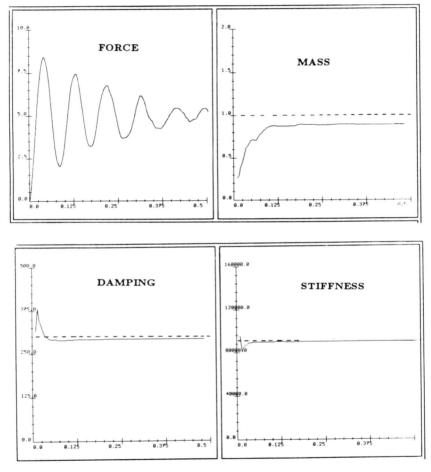

Figure 8.9: Simulation of estimation of environment dynamics.

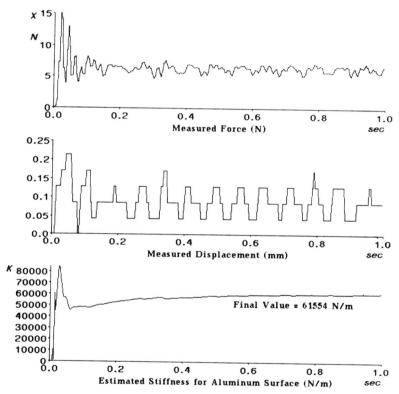

Figure 8.10: Experimental stiffness estimate on aluminum surface.

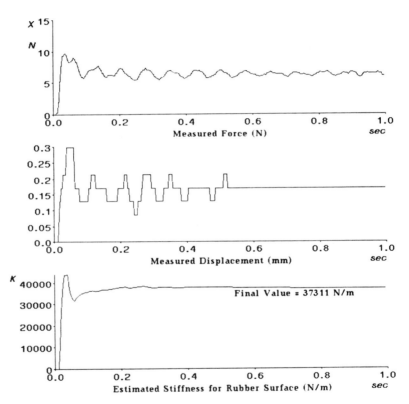

Figure 8.11: Experimental stiffness estimate on rubber surface.

attached to the sensor) together is lower than the actual stiffness. In robot control, it is this effective stiffness that is important for stability considerations. The estimates for the mass and the damping were not consistent and are not shown. As discussed in the simulation section, this was expected since those estimates depend on the derivatives of the position data. For the examples presented, the positional displacements are too small for the derivatives to be accurate. With the 16 bit resolution resolver, the maximum displacements are only 2.5 bits for aluminum and 3.5 bits for rubber.

Other Methods

There are other methods for identifying the environment dynamics. An adaptive observer technique (Narendra and Kudva, 1974; Narendra and Valavani, 1976; Shih, 1985) may be useful. Another possible method is to estimate the environment stiffness k_E by estimating the frequency of oscillation caused by interaction with the environment. The frequency information may be obtained by taking the FFT of the force measurement f_s.

8.3.3 Feasibility

Though theoretically possible to identify the dynamics of the environment, the experimental results indicate that there is not enough resolution in the joint angle sensor to measure x_E accurately. Given this measurement constraint, practical estimation limits in the stiffness of an environment can be computed by the following simple static analysis:

$$k_E(limit) = f/x_E$$

where f is the acceptable interaction force at the tip and x_E is the resolution of the position measurement at the tip. Assuming that the force sensor is much stiffer than the environment, that the joint angle can be measured to 16 bits (0.0055°), and that the lever arm is 0.5 m, the estimation limits are:

f	k_E
$1N$	$2 \times 10^4 \, N/m$
$10N$	$2 \times 10^5 \, N/m$

Unless the contact force is undesirably large, stiff environments cannot be identified accurately. This characteristic will limit the usefulness of this approach in dealing with stiff environments.

8.4 Joint Torque Control

Since very high-gain position feedback results from a high force feedback gain and a stiff environment, stability can be improved by lowering the force feedback gain. Unfortunately the force resolution would deteriorate. Another approach achieves force control without the tip force sensor in the feedback loop, by relying on the measurements of and the ability to command joint torques accurately. The loop gain is zero with this method, which is an open-loop control from the point of view of interaction forces at the end effector. In actuality, a closed-loop torque servo is implemented at each joint so that the joint torque can be specified accurately.

This approach was investigated previously in the context of position control by (Wu and Paul, 1980; Luh, Fisher, and Paul, 1983). Strain gauges were placed on the motor shaft of the geared joints of the Stanford Arm, and joint torque feedback was used mainly to reduce frictional effects at the joints. The experimental joint torque control method was not used to perform active force control. For the DDArm, since there is very little friction at the joints, the measurements of joint torques can be obtained by measuring the motor currents.

Wu and Paul (1980) presented a good comparison between wrist force sensing and joint torque sensing in implementing force control. Wrist sensing provides accurate force/torque measurements at the hand; but because the robot structure is inherently a low bandwidth flexible system and the sensor is situated at the end of this structure, a high gain feedback will produce instability as shown in Section 8.1. Therefore, only a slow closed loop system can be implemented stably using a wrist sensor. On the other hand, since joint torque sensors are situated before the low bandwidth robot structure, a high bandwidth torque inner loop can be implemented around each joint. Since the sensors are not at the hand, the hand forces and torques cannot be inferred as accurately as by wrist sensing.

Since both wrist sensing and joint sensing have good and bad features, one reasonable method is to combine the two methods to provide a stable high bandwidth force-controlled system. The high bandwidth open-loop joint torque control with an inner torque servo loop will provide stability and fast response, and the lower bandwidth outer loop with wrist force sensing will provide steady state accuracy. This method does not appear to have been implemented before, perhaps due to the lack of a suitable manipulator. The accurate joint torque control of the DDArm readily

Dynamic Stability Issues in Force Control

lends itself to this implementation.

In the rest of this section, the stability of an open-loop force controller using joint torque sensing is discussed in more detail. The results of one-link force control experiments are presented, demonstrating the performance of joint torque sensing and also the combination of joint torque and wrist force sensing.

8.4.1 Dominant Pole

Without the wrist sensor in the force feedback loop, the dynamics of a simple manipulator in contact with its environment is given by

$$f - k_E x = m\ddot{x} + b\dot{x} \tag{8.19}$$

If this system is controlled purely by commanding the force or the torque at the joint, i.e. in open loop mode, the response should be very underdamped since k_E may be high for a stiff environment. One simple classical compensation method is to create a dominant pole in the loop transfer function (Roberge, 1975). This can be done simply by putting a low-pass filter in the forward path so that

$$\tau = f \frac{a}{s+a}, \tag{8.20}$$

where τ is the actual input torque (or currents) to the actuator and a is much less than the resonant frequency of the original system. The total system dynamics is described by

$$x(s) = f \left(\frac{a}{s+a}\right) \left(\frac{1}{ms^2 + bs + k_E}\right). \tag{8.21}$$

Because of the dominant pole at $s = a$, this compensated system behaves in a much more stable manner despite the two high frequency underdamped poles. Figure 8.12 compares the step responses of the above system (8.21) to the original system without the dominant pole (8.19) for $m = 1, b = 10, k_E = 10000$, and $a = 5$.

Real actuators may obviate the need for explicit low-pass filtering of the input signal, since the amplifiers plus manipulator structure behave essentially as low-pass filters. The DDArm has an analog current loop at each motor amplifier which, together with the rotor inertia, has a cutoff frequency at approximately 30 Hz. Hence explicit low-pass filtering is unnecessary. Unfortunately, as discussed in Chapter 2, even direct

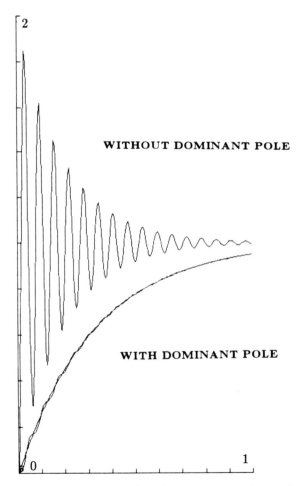

Figure 8.12: Step responses of (8.19) with and without the dominant pole.

Dynamic Stability Issues in Force Control

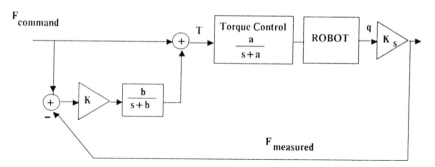

Figure 8.13: Structure of a general force controller.

drive motors have nonlinear characteristics, such as deadzones for small torques, cogging, and imperfect commutation circuitry. Some of these nonlinear characteristics were reduced by implementing another torque feedback loop at each joint (Figure 2.6).

Though force control via joint torque sensing is stable and well behaved, it is also desirable to include the wrist force sensing to improve steady state accuracy. Analogously, the force sensor feedback should be low-pass filtered to create a dominant pole, but at a much lower frequency than the pole of the joint torque path, so that the stability is not affected. The combined multi-feedback loop system should have stability and high bandwidth from the joint torque control mechanism, and steady state accuracy from the wrist force control mechanism. The block diagram of this system is shown in Figure 8.13.

8.4.2 One Link Force Control Experiments

Stable force controllers with and without the wrist force sensor were implemented on the third link of the DDArm, and are summarized below.

1. Using joint torque sensing only:

$$\tau = f_{desired} \cdot l_3, \quad l_3 = 0.4445\,m$$

2. Combination of joint torque sensing and wrist force sensing:

$$\tau = f_{desired} \cdot l_3 + k_f \left(\frac{b}{s+b}\right)(f_{desired} - f_{measured})$$

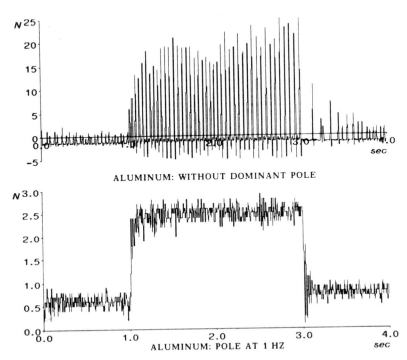

Figure 8.14: Force step responses with and without the dominant pole.

The third link has an explicit joint torque feedback loop shown in Figure 2.6 implemented digitally at $500\,Hz$. In the implementation using joint torque sensing only, the wrist force sensor is used only to record data.

Force Step Response

For the robot in contact with a stiff aluminum surface, Figure 8.14 compares the second method to a simple pure gain force feedback without the dominant pole. As before, the negative bias shown in the top plot is from an offset drift in the force sensor and should be ignored. The plots show raw data from the force sensor, without any low-pass filtering of the data to reduce the noise. Although there is a significant amount of noise in the force data, the step response of the compensated system is definitely stable, whereas the response of the system without proper compensation is unstable and the manipulator bounces on the environment surface.

Dynamic Stability Issues in Force Control

Figure 8.15 shows the force step responses of the two stable implementations discussed above. As expected from the analysis, there is no noticeable difference in the dynamic behavior between the two controllers. Yet the second method, combining the joint torque sensing and the wrist force sensing, has much better steady state accuracy. For either method, the accuracy is worse for the lower force inputs since the nonlinear characteristics (such as deadzone and cogging) of the motor and the amplifier for small torque levels are more severe. The commanded force levels ($10\,N$ and $15\,N$) in the experiments are less than 3% of the capacity of the DDArm, which can exert greater than $500\,N$ of force at its tip (but you don't want to be anywhere near it when it does).

Performance Tests

Steps are not really the best inputs in testing the performance of force control methods, although stability is qualitatively tested. Better performance measures are command following and disturbance rejection. These can be tested by the following inputs respectively:

1. sine wave force command with the manipulator in contact with the stationary environment, or

2. constant force command with the manipulator in contact with the environment moving in a sinusoidal trajectory.

Both of these test inputs give information about bandwidth. In the following, only the results for the controller with both joint torque and wrist force sensing are presented. As in the step response results, experiments with only joint torque sensing showed less accuracy without much difference in the dynamic behavior.

Sine Wave Force Command. The responses to several sine wave force commands are given in Figure 8.16. Neglecting the bias errors, the manipulator follows the $1Hz$ sine wave command faithfully. Although there is a noticeable lag in the measured data for higher frequencies, the manipulator still has no trouble following the sine wave commands. From the usual definition of bandwidth (frequency at which the output magnitude is reduced by 3dB from the DC value), the bandwidth tested in this way is greater than $20Hz$.

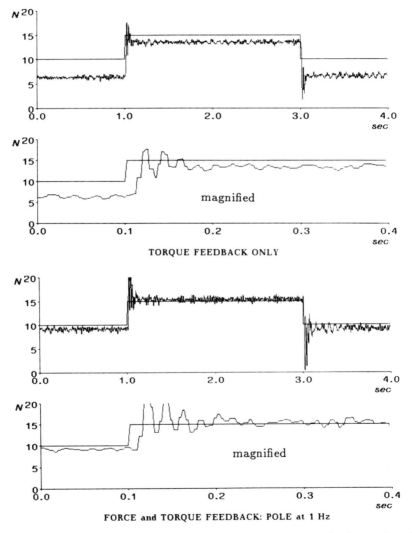

Figure 8.15: Force step responses for the two stable methods overlayed on the commanded step inputs.

Dynamic Stability Issues in Force Control 163

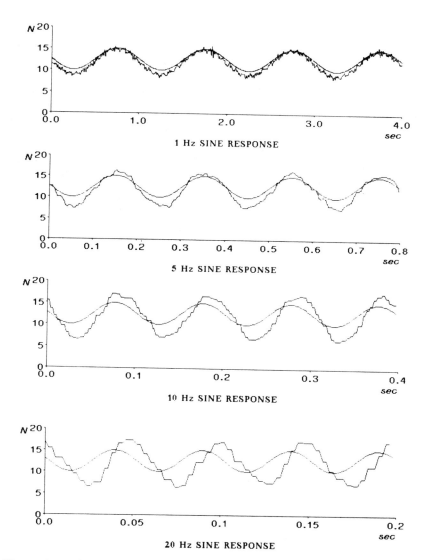

Figure 8.16: Sine wave responses for the second method using both joint torque and wrist force sensing.

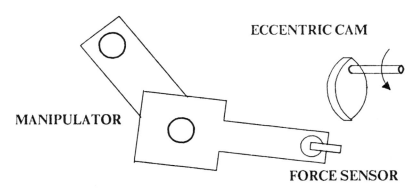

Figure 8.17: Setup for force response to positional disturbance.

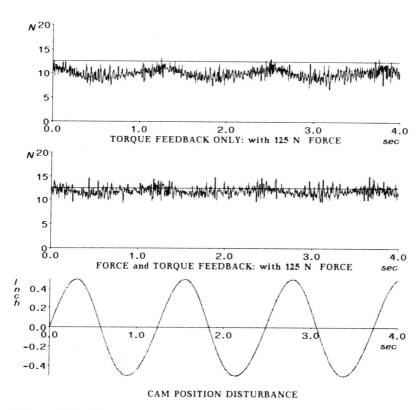

Figure 8.18: Force responses to a sinusoidal positional disturbance.

Constant Force With Positional Disturbances. The second performance test, a constant force command while the contacted environment is moving, is made with an eccentric cam on a gear motor (Figure 8.17). The circular cam is attached to the gear motor at a point $0.5\,in.$ off-center, providing an approximate sine wave positional disturbance to the force controller. The responses are plotted in Figure 8.18. The manipulator is able to follow the turning cam accurately with the desired force. Neglecting the high frequency noise, the low frequency variation in the measured force is within 10% of the desired $12.5N$. Higher cam speeds resulted in larger variations.

A reasonable measure of this disturbance rejection is the frequency w_{dr} at which the variation is 10%.

$$\left.\frac{f_{command} - f_s}{f_{command}}\right|_{w_{dr}} = 10\% \tag{8.22}$$

Under this measure, this system is well behaved up to approximately $0.8Hz$. The frequency limit determined by the positional disturbance is much lower than the bandwidth determined by the sinusoidal force trajectory testing. This is not surprising since for the sinusoidal trajectory tests, the movement of the link is infinitesimal, whereas for the cam tests, the controller has to move the large link inertia over a significant distance. It is not clear which measurement of performance is more useful. For applications that involve following undulated surfaces, the positional disturbance test is more relevant. For peg-in-hole applications with small movements, the sinusoidal trajectory test may be more relevant.

8.5 Discussion

Dynamic stability issues in the force control of manipulators were analyzed in this chapter by simple models and by implementation on the third link of the DDArm. In accord with past observations, it was shown that simple force control with tip force sensor feedback is unstable during contact with stiff environments. The root of the instability is that the environmental stiffness effectively multiplies the force feedback gain, leading to a highly underdamped system. Unmodeled dynamics and noise can easily drive the system unstable.

We first examined two approaches to stable force control, compliant coverings and adaptation to the environmental dynamics. The rationale for compliant coverings is to dominate the stiffness of the environment by the compliance of the covering or of a soft sensor. Drawbacks of compliant coverings or soft sensors include loss of dynamic range and position accuracy. Adaptation to environmental dynamics is theoretically attractive, but in reality unpractical. Position sensing is insufficiently accurate to estimate the environmental dynamics for stiff environments.

We proposed to achieve stability instead by implementing force control through essentially open-loop joint torque control. A low-pass filtered tip force sensor feedback was added to increase steady state accuracy without affecting stability. Experimental results on the direct drive arm showed that this method can accurately and stably exert forces on stiff environments.

Chapter 9

Kinematic Stability Issues in Force Control

This chapter identifies a new and surprising form of instability in the force control of manipulators. One form of instability, due to hard contact conditions, occurs in single-joint or multi-joint systems and was discussed in Chapter 8. The new form of instability that we have identified occurs only in multi-joint manipulators and is due to an inverse kinematic transformation in the feedback path. This instability is present only in certain formulations of the simultaneous force and position control of manipulators, and then only for manipulators with certain kinematic structures.

A variety of force control schemes have been devised based on Cartesian coordinates of the endpoint or of an external reference frame. These include hybrid control (Raibert and Craig, 1981), stiffness control (Salisbury, 1980), resolved acceleration control (Luh, Walker and Paul, 1980b; Shin and Lee, 1985), operational space method (Khatib, 1983), and impedance control (Hogan, 1985a-c; Kazerooni, Sheridan, and Houpt, 1986a-c). The rationale for Cartesian position/force control is that the geometry of the external world defines a set of natural coordinates that can be partitioned into position-controlled variables versus force-controlled variables (Mason, 1981; Lipkin and Duffy, 1986). Control, therefore, is cast in terms of these variables, necessitating a kinematic transformation to joint angles.

The major classes of Cartesian force control schemes are shown in Figure 9.1, and are summarized below.

Hybrid Control. The Cartesian positions and the velocities are computed from the joint positions and velocities, respectively, by direct or forward kinematics (Raibert and Craig, 1981). Neglecting the integral terms,

$$\tau = \mathbf{K}_{pj}\mathbf{J}^{-1}\mathbf{S}(\mathbf{x}_d - \mathbf{x}) + \mathbf{K}_{vj}\mathbf{J}^{-1}\mathbf{S}(\dot{\mathbf{x}}_d - \dot{\mathbf{x}}) + \mathbf{K}_f\mathbf{J}^T(\mathbf{I} - \mathbf{S})(\mathbf{f}_d - \mathbf{f}) \quad (9.1)$$

where $\mathbf{x}$ and $\mathbf{x}_d$ are the actual and desired Cartesian positions, $\mathbf{f}$ and $\mathbf{f}_d$ are the actual and desired external forces, $\mathbf{J}$ is the Jacobian matrix, $\mathbf{K}_{pj}$ and $\mathbf{K}_{vj}$ are the position and velocity gain matrices in joint coordinates, $\mathbf{I}$ is the identity matrix, and $\mathbf{S}$ is the diagonal selection matrix. The (i,i) entry of $\mathbf{S}$ is 1 if the ith axis is to be position controlled, and 0 if it is to be force controlled.

Stiffness Control. Proposed by Salisbury (1980), stiffness control was originally cast to compute kinematic errors in joint coordinates $\mathbf{q}$:

$$\tau = \mathbf{J}^T\mathbf{K}_p\mathbf{J}(\mathbf{q}_d - \mathbf{q}) + \mathbf{K}_{vj}(\dot{\mathbf{q}}_d - \dot{\mathbf{q}}) \quad (9.2)$$

where $\mathbf{K}_p$ is the stiffness matrix in Cartesian coordinates. By slightly modifying the stiffness control algorithm, errors can be computed in Cartesian coordinates as in hybrid control:

$$\tau = \mathbf{J}^T(\mathbf{K}_p(\mathbf{x}_d - \mathbf{x}) + \mathbf{K}_v(\dot{\mathbf{x}}_d - \dot{\mathbf{x}})) \quad (9.3)$$

Since the stiffness, not the pure force, is to be controlled, the above controller equations are shown without any force feedback term. A force term, however, can be added if the stiffness matrix alone does not provide enough force resolution or if pure force control is desired in some direction.

Resolved Acceleration. Although Luh, Walker, and Paul (1980b) originally formulated resolved acceleration for position control only, it can be simply reformulated to control force and position simultaneously (Shin and Lee, 1985). The modified resolved acceleration controller is:

$$\tau = \mathbf{H}\mathbf{J}^{-1}[\mathbf{S}\mathbf{x}_m^* - \dot{\mathbf{J}}\dot{\mathbf{q}}] + \dot{\boldsymbol{\theta}} \cdot \mathbf{C} \cdot \dot{\boldsymbol{\theta}} + \mathbf{g} + \mathbf{J}^T(\mathbf{I} - \mathbf{S})\mathbf{f}^* \quad (9.4)$$

where

$$\mathbf{x}_m^* = \ddot{\mathbf{x}}_d + \mathbf{K}_v(\dot{\mathbf{x}}_d - \dot{\mathbf{x}}) + \mathbf{K}_p(\mathbf{x}_d - \mathbf{x}). \quad (9.5)$$

$\mathbf{f}^*$ is the command vector for active force control, which is the only modification from the original formulation.

The modified resolved acceleration control heads a class of essentially identical methods, including the operational space method (Khatib, 1983)

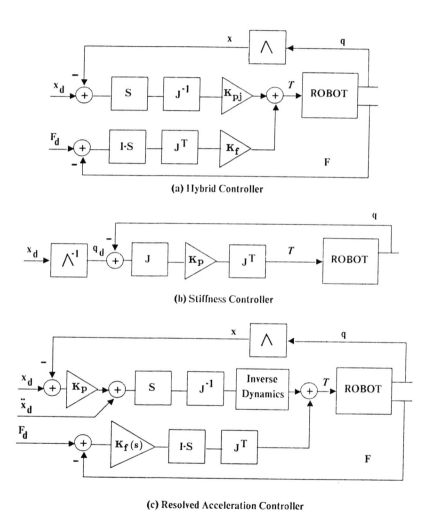

Figure 9.1: Block diagrams for the hybrid, the stiffness, and the resolved acceleration controllers shown without the velocity terms.

and impedance control (Hogan, 1985a-c). The equivalence of the operational space method is shown in Appendix 4; it was also shown in (DeSchutter, 1986). For impedance control, the only difference is that there is an additional force feedback term to alter the apparent inertia of the robot.

Actually stiffness control and resolved acceleration are also hybrid controllers, in that they can be formulated to control position and force for different axes. For convenience, when the term hybrid control is mentioned in this chapter without any other clarifying description, we are always referring to the method of Raibert and Craig.

These methods differ in how the kinematic transformation is accomplished and in whether the dynamics of the manipulator are incorporated. These differences will turn out to be crucial in determining stability, again emphasizing the importance of a model-based control. In hybrid control and stiffness control, the dynamic model of the robot is not included, whereas in resolved acceleration control it is. The inverse Jacobian appears in hybrid control and in resolved acceleration control, but only the Jacobian transpose appears in stiffness control.

It will be shown that stiffness control and resolved acceleration control do not become unstable for either revolute or polar manipulators. Hybrid control, on the other hand, is inherently unstable for revolute manipulator regardless of the choice of gains, although it is stable for polar manipulators. These instabilities occur not only at the points of kinematic singularities, where the Jacobian inverses are not defined, but at a wide range of the manipulator work space, where the Jacobian inverses are well defined.

This chapter includes both analytical and experimental results with the DDArm. Indeed, the motivation for this work came from experimental observations of instability when the hybrid controller was tried on the DDArm.

9.1 Intuitive Stability Analysis

Before a more rigorous treatment, it is helpful to get an intuitive understanding of the source of the kinematic instability. The analysis presented in this section, which covers just hybrid and stiffness control, is approximate and *not* mathematically rigorous; a more rigorous analysis is presented in the next section and the discussion of resolved acceleration control is postponed until that section. The analysis will be applied to

Kinematic Stability Issues in Force Control

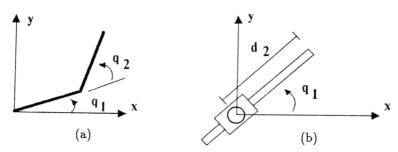

Figure 9.2: (A) Simple two-joint rotary planar manipulator. (B) Simple planar polar manipulator.

both two-joint planar rotary and polar manipulators (Figure 9.2). It will be shown that the instability occurs only for the rotary planar manipulator.

9.1.1 Hybrid Control

The equation for the hybrid controller (9.1) is repeated below:

$$\tau = \mathbf{K}_{pj}\mathbf{J}^{-1}\mathbf{S}\mathbf{x}_e + \mathbf{K}_{vj}\mathbf{J}^{-1}\mathbf{S}\dot{\mathbf{x}}_e + \mathbf{K}_f\mathbf{J}^T(\mathbf{I}-\mathbf{S})(\mathbf{f}_d - \mathbf{f}) \tag{9.6}$$

where

$$\mathbf{x}_e = \mathbf{x}_d - \mathbf{x}$$

$$\mathbf{S} = \begin{bmatrix} 1 & 0 \\ 0 & 1 \end{bmatrix} \text{ or } \begin{bmatrix} 0 & 0 \\ 0 & 1 \end{bmatrix} \text{ or } \begin{bmatrix} 1 & 0 \\ 0 & 0 \end{bmatrix} \text{ or } \begin{bmatrix} 0 & 0 \\ 0 & 0 \end{bmatrix}$$

The first $\mathbf{S}$ specifies position control in both the x and y directions. The second $\mathbf{S}$ specifies position control only in the y direction and force control in the x direction, and the reverse for the third $\mathbf{S}$. The fourth $\mathbf{S}$ specifies pure force control in both directions.

The Two-joint Planar Rotary Manipulator

For this rotary manipulator, the joint coordinates are $\mathbf{q} = \boldsymbol{\theta} = (\theta_1, \theta_2)$. The kinematics are:

$$\begin{bmatrix} x \\ y \end{bmatrix} = \begin{bmatrix} l_1 c_1 + l_2 c_{12} \\ l_1 s_1 + l_2 s_{12} \end{bmatrix} \tag{9.7}$$

$$\begin{bmatrix} \dot{x} \\ \dot{y} \end{bmatrix} = \begin{bmatrix} -l_1 s_1 - l_2 s_{12} & -l_2 s_{12} \\ l_1 c_1 + l_2 c_{12} & l_2 c_{12} \end{bmatrix} \begin{bmatrix} \dot{\theta}_1 \\ \dot{\theta}_2 \end{bmatrix} \quad (9.8)$$

$$\begin{bmatrix} \dot{\theta}_1 \\ \dot{\theta}_2 \end{bmatrix} = \frac{1}{l_1 l_2 s_2} \begin{bmatrix} l_2 c_{12} & l_2 s_{12} \\ -l_1 c_1 - l_2 c_{12} & -l_1 s_1 - l_2 s_{12} \end{bmatrix} \begin{bmatrix} \dot{x} \\ \dot{y} \end{bmatrix} \quad (9.9)$$

where $s_1 = \sin(\theta_1)$ and $s_{12} = \sin(\theta_1 + \theta_2)$. In the simulations, $l_1 = 0.462m$ and $l_2 = 0.4445m$, representing the lengths of the DDArm configured as a two-link planar manipulator by locking the second joint and moving only the first and the third joints.

Suppose the manipulator is in free space and has no mass at the end of the force sensor. Then for any axis selected to be force controlled, the desired and the measured forces are 0's. From the dynamic stability point of view of Chapter 8, this situation is the most stable. Consider just the position component, since the velocity component is analogous, and let $\mathbf{K}_{pj} = \mathbf{I} > 0$ for simplicity. Then

$$\boldsymbol{\tau} = \mathbf{K}_{pj}\mathbf{J}^{-1}\mathbf{S}\mathbf{x}_e = \mathbf{J}^{-1}\mathbf{S}\mathbf{x}_e \approx \mathbf{J}^{-1}\mathbf{S}\mathbf{J}\boldsymbol{\theta}_e \quad (9.10)$$

where $\boldsymbol{\theta}_e = \boldsymbol{\theta}_d - \boldsymbol{\theta}$.

Case 1. Let $\mathbf{S} = \mathbf{I}$, i.e. position control in both x and y directions. Then

$$\boldsymbol{\tau} = \mathbf{J}^{-1}\mathbf{S}\mathbf{J}\boldsymbol{\theta}_e = \boldsymbol{\theta}_e$$

In this case, the hybrid controller becomes a simple independent-joint controller and the system remains stable assuming that the control designer has designed a stable joint control system.

Case 2. Let $\mathbf{S} = \text{diag}(0, 1)$, i.e. position control in y and force control in x. Since the manipulator is assumed to be in free space, the x axis force is being controlled to zero. With this selection matrix, position errors in the x direction should not cause any restoring torques. Expanding the simplified hybrid controller equation,

$$\boldsymbol{\tau} = \frac{1}{l_1 l_2 s_2} \begin{bmatrix} l_2 s_{12}(l_1 c_1 + l_2 c_{12}) & l_2 s_{12}(l_2 c_{12}) \\ (-l_1 s_1 - l_2 s_{12})(l_1 c_1 + l_2 c_{12}) & (-l_1 s_1 - l_2 s_{12})(l_2 c_{12}) \end{bmatrix} \begin{bmatrix} \theta_{1e} \\ \theta_{2e} \end{bmatrix}. \quad (9.11)$$

Consider the configuration with $\theta_1 \approx 0°$, $0° < \theta_2 < 90°$, $\theta_{1e} = 0°$, and $\theta_{2e} > 0°$ (Figure 9.3A). Although θ_{2e} has both x and y components, the

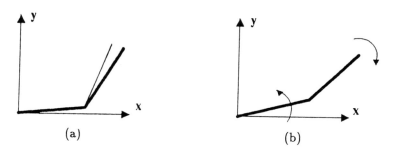

Figure 9.3: (A) Unstable configuration for hybrid control. (B) Hybrid control response with $\mathbf{S} = \text{diag}(0, 1)$.

hybrid controller is being commanded to correct for the y-axis error only. From (9.11) the restoring torques are:

$$\tau_1 = [\frac{1}{l_1 l_2 s_2}(l_2 s_{12})(l_2 c_{12})]\theta_{2e} > 0$$

$$\tau_2 = -[\frac{1}{l_1 l_2 s_2}(l_1 s_1 + l_2 s_{12})(l_2 c_{12})]\theta_{2e} < 0$$

For τ_2, the quantity inside the bracket is positive for the configuration considered above, and hence there is positive feedback about joint 2. The response is shown in Figure 9.3B; τ_1 is positive and τ_2 is negative, and the manipulator opens up, moves through a singularity, and becomes unstable.

Since the x axis is being force controlled to $f_x = 0$, and there is no restoring force to the position error in the x axis, this response seems to be an admissible response if only seen from the point of view of Cartesian coordinates. The restoring torques are moving the manipulator to reduce the position error in the y direction without any concern for position error in the x direction. In reality, this behavior is unstable since the manipulator is being pushed toward a singularity, where $\mathbf{J}^{-1}$ is undefined.

The Two-joint Planar Polar Manipulator

For the polar manipulator, the joint coordinates are $\mathbf{q} = (\theta_1, d_2)$. The kinematics are:

$$\begin{bmatrix} x \\ y \end{bmatrix} = \begin{bmatrix} d_2 \cos \theta_1 \\ d_2 \sin \theta_1 \end{bmatrix} \quad (9.12)$$

$$\begin{bmatrix} \dot{x} \\ \dot{y} \end{bmatrix} = \begin{bmatrix} -d_2 \sin\theta_1 & \cos\theta_1 \\ d_2 \cos\theta_1 & \sin\theta_1 \end{bmatrix} \begin{bmatrix} \dot{\theta}_1 \\ \dot{d}_2 \end{bmatrix} \quad (9.13)$$

$$\begin{bmatrix} \dot{\theta}_1 \\ \dot{d}_2 \end{bmatrix} = -\frac{1}{d_2} \begin{bmatrix} \sin\theta_1 & -\cos\theta_1 \\ -d_2 \cos\theta_1 & -d_2 \sin\theta_1 \end{bmatrix} \begin{bmatrix} \dot{x} \\ \dot{y} \end{bmatrix} \quad (9.14)$$

Following the intuitive analysis for the revolute manipulators, when $\mathbf{S} = \text{diag}(0,1)$,

$$\tau = \frac{1}{2} \begin{bmatrix} d_2 \cos^2\theta_1 & \cos\theta_1 \sin\theta_1 \\ d_2^2 \sin\theta_1 \cos\theta_1 & d_2 \sin^2\theta_1 \end{bmatrix} \begin{bmatrix} \theta_{1e} \\ d_{2e} \end{bmatrix} \quad (9.15)$$

The diagonal elements of the above gain matrix are always positive, resulting in stable negative feedbacks if, as before, the nonlinear coupled system were to be approximated as two independent SISO systems. Therefore, we would expect polar manipulators not to exhibit kinematic instabilities under hybrid control.

9.1.2 Stiffness Control

The previously demonstrated kinematic instability does not occur with stiffness control. Consider again only the position-dependent stiffness component of the controller and assume that the manipulator is in free space. For the two-joint rotary manipulator,

$$\tau = \mathbf{J}^T \mathbf{K}_p \mathbf{J} (\boldsymbol{\theta}_d - \boldsymbol{\theta}) = \mathbf{J}^T \mathbf{K}_p \mathbf{J} \boldsymbol{\theta}_e. \quad (9.16)$$

Assuming that the stiffness matrix $\mathbf{K}$ is diagonal, the above equation is expanded as

$$\begin{aligned} \mathbf{J}^T \mathbf{K}_p \mathbf{J} \boldsymbol{\theta}_e &= \begin{bmatrix} J_{11} & J_{21} \\ J_{12} & J_{22} \end{bmatrix} \begin{bmatrix} k_x & 0 \\ 0 & k_y \end{bmatrix} \begin{bmatrix} J_{11} & J_{12} \\ J_{21} & J_{22} \end{bmatrix} \begin{bmatrix} \theta_{1e} \\ \theta_{2e} \end{bmatrix} \quad (9.17) \\ &= \begin{bmatrix} k_x J_{11}^2 + k_y J_{21}^2 & k_x J_{11} J_{12} + k_y J_{21} J_{22} \\ k_x J_{11} J_{12} + k_y J_{21} J_{22} & k_x J_{12}^2 + k_y J_{22}^2 \end{bmatrix} \begin{bmatrix} \theta_{1e} \\ \theta_{2e} \end{bmatrix}. \end{aligned}$$

The diagonal entries of $\mathbf{J}^T \mathbf{K}_p \mathbf{J}$ are always positive for positive stiffness values. For the same manipulator configuration as before, where only the y axis is being position controlled ($k_x = 0$):

$$\begin{aligned} \tau_1 &= k_y J_{21} J_{22} = k_y (l_1 c_1 + l_2 c_{12}) + l_2 c_{12} > 0 \\ \tau_2 &= k_y J_{22}^2 > 0 \end{aligned}$$

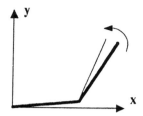

Figure 9.4: Response of the stiffness Controller with $\mathbf{S} = \text{diag}(0,1)$.

Unlike hybrid control, τ_2 is positive, which is indicative of negative feedback about joint 2. As shown in Figure 9.4, the manipulator corrects for the error in y position without moving toward a singularity. This stable behavior can be generalized by simple matrix operations for manipulators with more than two joints and for other manipulator configurations and parts of the workspace.

9.2 Root Loci, Simulations, and Experiments

The analyses of the previous section suggest that hybrid control is potentially unstable, due to the Jacobian inverse in the coordinate transformation. The Jacobian transpose in stiffness control does not seem to create a similar problem. Nevertheless, as stated at the beginning of that section, the intuitive analyses are only approximate. A typical manipulator is a full MIMO system with the dynamics of each state coupled to one another. However, we treated the multivariable manipulator system essentially as SISO systems, and looked at the responses due to an error in only one component of the state vector.

In this section, the intuition gained from the previous approximate analyses is verified by more rigorous stability analyses and experiments. A Lyapunov method can be applied for global stability analysis; but since the purpose of this chapter is to show that some well-accepted force controllers are unstable, it is sufficient to study local stability by computing the closed-loop eigenvalues of the linearized manipulator system about some equilibrium point. The eigenvalues are evaluated for different force controllers, and the results are verified by experiments on the DDArm. For the case of polar manipulators, the stability results are verified by simulations only.

The rigid body dynamics of a manipulator without gravity is given by:

$$\tau = \mathbf{H}(\mathbf{q})\ddot{\mathbf{q}} + \dot{\mathbf{q}} \cdot \mathbf{C}(\mathbf{q}) \cdot \dot{\mathbf{q}} \qquad (9.18)$$

Assuming negligible velocities ($\dot{\mathbf{q}} = 0$) and linearizing this equation about some nominal position $\mathbf{q}$,

$$\delta\tau = \mathbf{H}(\mathbf{q})\delta\ddot{\mathbf{q}}. \qquad (9.19)$$

In state space form,

$$\delta\dot{\mathbf{x}} = \begin{bmatrix} \mathbf{0} & \mathbf{I} \\ \mathbf{0} & \mathbf{0} \end{bmatrix} \delta\mathbf{x} + \begin{bmatrix} \mathbf{0} \\ \mathbf{H}(\mathbf{q})^{-1} \end{bmatrix} \delta\tau, \qquad (9.20)$$

where $\delta\mathbf{x} = (\delta\mathbf{q}, \delta\dot{\mathbf{q}})$.

9.2.1 Hybrid Control

In state space form, the hybrid controller can be written as

$$\delta\tau = \begin{bmatrix} -\mathbf{K}_{pj}\mathbf{J}^{-1}\mathbf{SJ}, & -\mathbf{K}_{vj}\mathbf{J}^{-1}\mathbf{SJ} \end{bmatrix} \begin{bmatrix} \delta\mathbf{q} \\ \delta\dot{\mathbf{q}} \end{bmatrix}. \qquad (9.21)$$

The closed-loop system is then described as

$$\begin{aligned} \delta\dot{\mathbf{x}} &= \begin{bmatrix} \mathbf{0} & \mathbf{I} \\ -\mathbf{H}^{-1}\mathbf{K}_{pj}\mathbf{J}^{-1}\mathbf{SJ} & -\mathbf{H}^{-1}\mathbf{K}_{vj}\mathbf{J}^{-1}\mathbf{SJ} \end{bmatrix} \delta\mathbf{x} \\ &= \mathbf{A}\delta\mathbf{x} \end{aligned} \qquad (9.22)$$

To guarantee local stability at the equilibrium points, the eigenvalues of $\mathbf{A}$ must have negative real parts.

The Two-joint Planar Rotary Manipulator

First we consider the stability of the rotary manipulator under hybrid control for the same configurations as in Section 9.1.1. As before, the robot is in free space, which is the most dynamically stable configuration. It will be shown later that contacts do not improve the kinematic instabilities.

The inertia matrix $\mathbf{H}(\boldsymbol{\theta})$ of the rotary manipulator is given by (Brady et al., 1982):

Kinematic Stability Issues in Force Control

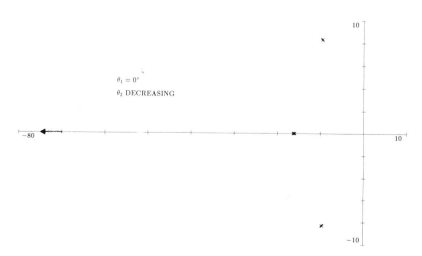

Figure 9.5: Root locus under hybrid control with $\mathbf{S} = \mathbf{I}$.

$$\mathbf{H}(\boldsymbol{\theta}) = \begin{bmatrix} h_1 + h_2 + m_2 l_1 l_2 \cos\theta_2 + \frac{1}{4}(m_1 l_1^2 + m_2 l_2^2) + m_2 l_1^2, & h_{21} \\ h_2 + \frac{1}{4}m_2 l_2^2 + \frac{1}{2}m_2 l_1 l_2 \cos\theta_2, & h_2 + \frac{1}{4}m_2 l_2^2 \end{bmatrix}$$

where m_i and h_i are the mass and inertia about the center of gravity of link i. The estimated inertial parameters of the DDArm (Chapter 5) are used in the analysis. After converting the 3-link parameters to those of the simple 2-link configuration, the inertial parameters are: $h_1 = 8.095 kg \cdot m^2$, $h_2 = 0.253\ kg \cdot m^2$, $m_1 = 120.1\ kg$, and $m_2 = 2.104\ kg$.

Case 1: $\mathbf{S} = \mathbf{I}$. Both the x and y axes are position controlled. The intuitive analysis showed that the hybrid control is stable in this case. The root locus diagram (Figure 9.5) shows the closed-loop eigenvalues of the hybrid controlled manipulator as θ_2 varies from $10°$ to $90°$ with $\theta_1 = 0°$. As expected, the eigenvalues are on the left hand side of the s-plane and the manipulator is stable.

Experimental results are shown in Figure 9.6. The manipulator is initially at $(\theta_1, \theta_2) = (0°, 70°)$ and the desired point is given in Cartesian coordinates as $(x, y) = (0.462\ m, 0.4445\ m)$. The gains ($k_{p1} = 2500$, $k_{v1} = 300$, $k_{p2} = 400$, $k_{v2} = 30$) are chosen experimentally for overdamped response. These results on the direct drive arm also verify that this case is indeed stable under hybrid control. Also shown is the manipulator under

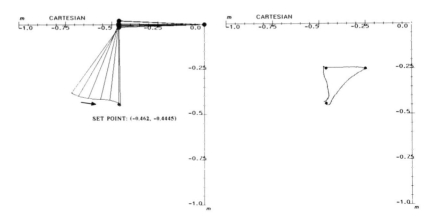

Figure 9.6: Experimental results under hybrid control with $\mathbf{S} = \mathbf{I}$.

hybrid control executing a triangular trajectory, defined by straight lines between the three corner points.

Case 2: $\mathbf{S} = \mathrm{diag}(0, 1)$. The x axis is force controlled to zero and the y axis is position controlled. The root locus of this case for $\theta_1 = 0$ and θ_2 varying from 90° to 70° is shown in Figure 9.7. Because one axis is being controlled to zero force, the behavior of the manipulator along that axis is of a pure mass or a double integrator. Therefore, two of the poles are at the origin, and only the remaining two poles are shown to be varying. For θ_2 near 90°, the system is stable and the eigenvalues are negative. Nevertheless, as θ_2 becomes smaller, the poles move into the right half of the s-plane. The interaction of the inertia matrix $\mathbf{H}$ with the $\mathbf{J}^{-1}$ matrix is such that the eigenvalues of the $\mathbf{A}$ matrix of the equation (9.22) have become positive.

In the root locus of Figure 9.7, the manipulator was assumed to be in free space. Figure 9.8 shows the root locus for the manipulator in contact with a stiff wall ($k_s = 10^5 \ N/m$). Since the instability is generated by the interaction of the inertia matrix and the Jacobian inverse, a manipulator in contact with its environment does not solve this instability problem.

Figure 9.9 shows the experimental results. The manipulator is initially at the stable position $(x, y) = (-0.462 \, m, -0.4445 \, m)$, according to the root locus diagram. Then, with a very light force, the tip of the manipulator is pulled along the $-x$ direction toward the more unstable configuration. The manipulator becomes unstable approximately at

Kinematic Stability Issues in Force Control

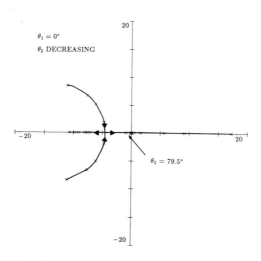

Figure 9.7: Root locus under hybrid control with $\mathbf{S} = \text{diag}(0,1)$.

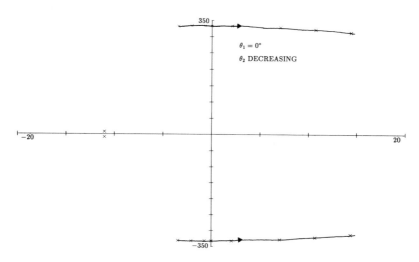

Figure 9.8: Root locus under hybrid control with $\mathbf{S} = \text{diag}(0,1)$ and in contact with a stiff wall ($k_s = 10^5 \ N/m$).

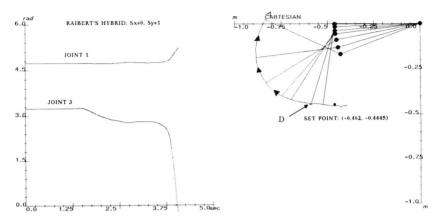

Figure 9.9: Experimental results under hybrid control with **S** = diag(0, 1).

$\theta_2 = 75°$ (point D), agreeing with the root locus diagram for the ideal manipulator showing positive poles for $\theta_2 < 79.5°$. Experiments with smaller feedback gains showed that the region of instability can be reduced, but that the inherent instability cannot be changed. Although we have not proved that this instability exists for all possible combinations of gains and friction, it seems that for the hybrid control, there will always be unstable regions encompassing the singularity points.

The Two-joint Planar Polar Manipulator

The dynamics are:

$$\begin{bmatrix} \tau_1 \\ f_2 \end{bmatrix} = \begin{bmatrix} (d_2 - \frac{1}{2}l_2)^2 m_2 & 0 \\ 0 & m_2 \end{bmatrix} \begin{bmatrix} \ddot{\theta}_2 \\ \ddot{d}_2 \end{bmatrix} + \begin{bmatrix} 2(d_2 - \frac{1}{2}l_2)m_2 \dot{d}_2 \dot{\theta}_1 \\ -(d_2 - \frac{1}{2}l_2)m_2 \dot{\theta}_1^2 \end{bmatrix} \quad (9.23)$$

where f_2 is the force required to move the sliding joint d_2.

The stability analysis of a polar manipulator is studied further by linearizing the dynamics about some equilibrium point as before. In order to be consistent with the results of Raibert and Craig (1981), the kinematic, dynamic, and control parameters reported in their study are used for evaluating hybrid control by root locus and simulations. The root locus with **S** = diag(0, 1) is shown in Figure 9.10 as d_2 varies from $0.05\,m$ to $0.6\,m$. The two non-zero eigenvalues are always negative in this

Kinematic Stability Issues in Force Control

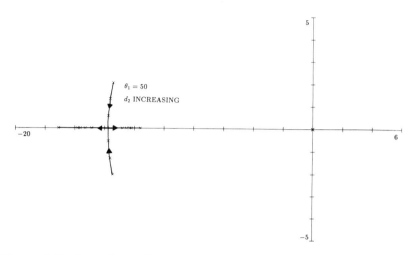

Figure 9.10: Root locus for hybrid control on a polar manipulator with $S = \text{diag}(0, 1)$.

case. The simulation shown in Figure 9.11 also demonstrates a stable behavior.

These results verify the stable results reported by Raibert and Craig on the Stanford Arm, which is a polar manipulator. Large friction in their manipulator may have also added stability to their implementations.

9.2.2 Resolved Acceleration Force Control

In state space form, the linearized resolved acceleration controller is written as

$$\delta\tau = \begin{bmatrix} -\hat{\mathbf{H}}\mathbf{J}^{-1}\mathbf{S}\mathbf{K}_p\mathbf{J}, & -\hat{\mathbf{H}}\mathbf{J}^{-1}\mathbf{S}\mathbf{K}_v\mathbf{J} \end{bmatrix} \begin{bmatrix} \delta\theta \\ \delta\dot{\theta} \end{bmatrix}. \tag{9.24}$$

$\hat{\mathbf{H}}$ is the model of the actual inertia matrix $\mathbf{H}$. The closed-loop system is described as

$$\delta\dot{\mathbf{x}} = \begin{bmatrix} 0 & \mathbf{I} \\ -\mathbf{H}^{-1}\hat{\mathbf{H}}\mathbf{J}^{-1}\mathbf{S}\mathbf{K}_p\mathbf{J} & -\mathbf{H}^{-1}\hat{\mathbf{H}}\mathbf{J}^{-1}\mathbf{S}\mathbf{K}_v\mathbf{J} \end{bmatrix} \delta\mathbf{x}. \tag{9.25}$$

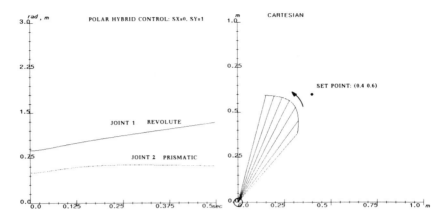

Figure 9.11: Simulation results for hybrid control on a polar manipulator with $\mathbf{S} = \text{diag}(0, 1)$.

If we assume perfect modeling so that $\hat{\mathbf{H}} = \mathbf{H}$, the closed-loop system becomes

$$\delta \dot{\mathbf{x}} = \begin{bmatrix} \mathbf{0} & \mathbf{I} \\ -\mathbf{J}^{-1}\mathbf{SK}_p\mathbf{J} & -\mathbf{J}^{-1}\mathbf{SK}_v\mathbf{J} \end{bmatrix} \delta \mathbf{x}$$
$$= \mathbf{A} \delta \mathbf{x}. \tag{9.26}$$

This $\mathbf{A}$ matrix can be decoupled as

$$\mathbf{A} = \begin{bmatrix} \mathbf{J}^{-1} & \mathbf{0} \\ \mathbf{0} & \mathbf{J}^{-1} \end{bmatrix} \begin{bmatrix} \mathbf{0} & \mathbf{I} \\ -\mathbf{SK}_p & -\mathbf{SK}_v \end{bmatrix} \begin{bmatrix} \mathbf{J} & \mathbf{0} \\ \mathbf{0} & \mathbf{J} \end{bmatrix}$$
$$= \mathbf{Q}^{-1}\mathbf{BQ}. \tag{9.27}$$

Thus $\mathbf{A}$ is a similarity transform of the stable matrix $\mathbf{B}$ consisting only of $\mathbf{S}$, $\mathbf{K}_p$, and $\mathbf{K}_v$. Since the eigenvalues of $\mathbf{B}$ are preserved under any similarity transform, the eigenvalues of $\mathbf{A}$ are stable by the choices of $\mathbf{K}_p$ and $\mathbf{K}_v$. Therefore, under perfect modeling, the resolved acceleration control results in a stable system. Unlike hybrid control, the inverse Jacobian matrix does not interact harmfully with the inertia matrix, which is cancelled in resolved acceleration control.

The eigenvalues are plotted in Figure 9.12 for the resolved acceleration controller with the following gains:

$$\mathbf{S} = \begin{bmatrix} 0 & 0 \\ 0 & 1 \end{bmatrix} \quad \mathbf{K}_p = \begin{bmatrix} 400 & 0 \\ 0 & 400 \end{bmatrix} \quad \mathbf{K}_v = \begin{bmatrix} 40 & 0 \\ 0 & 40 \end{bmatrix}$$

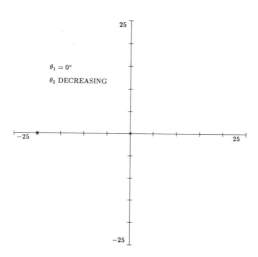

Figure 9.12: Root locus for resolved acceleration control with $\mathbf{S} = \text{diag}(0, 1)$.

As expected, the eigenvalues are negative and do not vary with the manipulator configuration since the matrix $\mathbf{B}$ in (9.27) is independent of the manipulator dynamics. This stability property is quite robust to modeling errors. Even with 50% error in inertial parameters, the eigenvalues remained in the left half plane. If the modeling errors are very large, then the eigenvalues will eventually become positive. For example, if we model the inertia matrix as the identity matrix, this reduces to the structure of the hybrid controller of Raibert and Craig, and the system will become unstable. Hence accurate modeling of the inertial terms of the manipulator is important in force control as well as in trajectory control. Since we only evaluated the stability for the cases when the velocity and the gravity terms are zero, we cannot conclusively state how these terms will affect kinematic stability.

Figure 9.13 shows the responses when both axes are position controlled. The response is stable, and for the same triangular trajectory as before, this controller follows the desired path much more accurately than the hybrid controller. This is expected since the controller structure includes the dynamics of the manipulator. The response when the y axis is position controlled and the x axis is force controlled to zero force is shown in Figure 9.14. The manipulator is stable and, as desired, only corrects for the error in the y direction, ignoring the position error in the

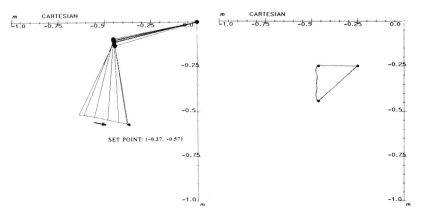

Figure 9.13: Experimental results for resolved acceleration control with $S = I$.

x direction.

9.2.3 Stiffness Control

The stiffness controller for the rotary manipulator is described in state space form as:

$$\delta\tau = \begin{bmatrix} -\mathbf{J}^T\mathbf{K}_p\mathbf{J}, & -\mathbf{J}^T\mathbf{K}_v\mathbf{J} \end{bmatrix} \begin{bmatrix} \delta\boldsymbol{\theta} \\ \delta\dot{\boldsymbol{\theta}} \end{bmatrix}. \qquad (9.28)$$

The closed-loop system is

$$\begin{aligned} \delta\dot{\mathbf{x}} &= \begin{bmatrix} \mathbf{0} & \mathbf{I} \\ -\mathbf{H}^{-1}\mathbf{J}^T\mathbf{K}_p\mathbf{J} & -\mathbf{H}^{-1}\mathbf{J}^T\mathbf{K}_v\mathbf{J} \end{bmatrix} \delta\mathbf{x} \\ &= \mathbf{A}\delta\mathbf{x} \end{aligned} \qquad (9.29)$$

Since dynamics are not included in the controller structure, the inertia matrix is not cancelled, and it is not obvious from the form of the **A** matrix what the stability characteristics should be. Consequently, stiffness control will be analyzed in the same way as hybrid controller.

Case 1: $\mathbf{K}_p = \text{diag}(2000, 2000)$. The root locus (Figure 9.15) for this case shows that the eigenvalues are always negative and the system is stable. The experimental results for this case are shown in Figure 9.16.

Kinematic Stability Issues in Force Control

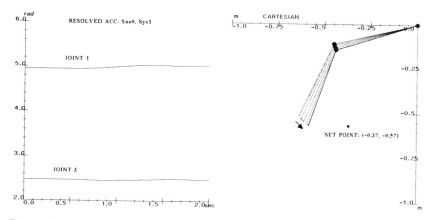

Figure 9.14: Experimental results for resolved acceleration control with $S = \text{diag}(0,1)$.

Since dynamics are not included, the stiffness controller performs poorly in following the triangular path. Yet, the response is always stable.

Case 2: $K_p = \text{diag}(0, 2000)$. The root locus (Figure 9.17) shows that this case is also stable as predicted by the intuitive analysis. The experimental results shown in Figure 9.18 also verifies the stability of this method.

9.3 Resolved Acceleration Force Control Experiments during Contact

To isolate kinematic instabilities from dynamic instabilities, the experiments in the previous section involved force controllers for the DDArm operating in free space without any interaction with the environment. In Chapter 8, one-link force control experiments demonstrated dynamic stability when the manipulator is in contact with a stiff environment. This section experimentally verifies the combined kinematic and dynamic stability with the resolved acceleration force control of the DDArm against a stiff environment.

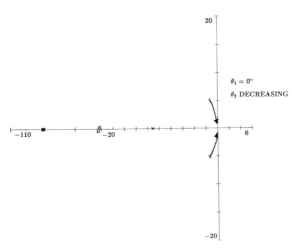

Figure 9.15: Root locus for stiffness control with $\mathbf{K}_p = \text{diag}(2000, 2000)$.

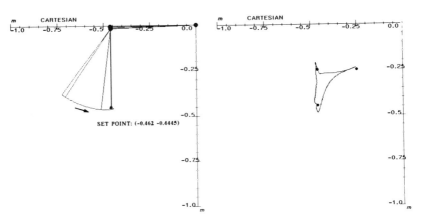

Figure 9.16: Experimental results for stiffness control with $\mathbf{K}_p = \text{diag}(2000, 2000)$.

Kinematic Stability Issues in Force Control

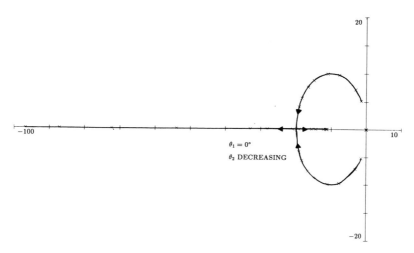

Figure 9.17: Root locus for stiffness control with $\mathbf{K}_p = \text{diag}(0, 2000)$.

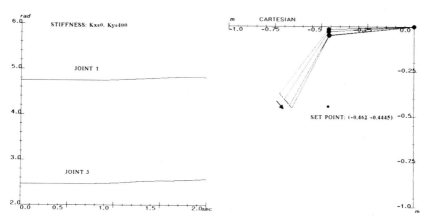

Figure 9.18: Experimental results for stiffness control with $\mathbf{K}_p = \text{diag}(0, 2000)$.

9.3.1 Experimental Setup

As in Section 9.2, the second joint of the 3-link DDArm is locked at $\theta_2 = 180°$ and the first and third joints are controlled to operate the arm in a planar configuration. The environment is a stiff aluminum surface. Dynamic stability is achieved with joint torque control and dominant pole compensated wrist force feedback. The general structure of the controller is:

$$\tau = \mathbf{F}(s)\{\mathbf{H}(\boldsymbol{\theta})\mathbf{J}^{-1}(\boldsymbol{\theta})[\mathbf{S}\mathbf{x}_m^* - \dot{\mathbf{J}}(\boldsymbol{\theta})\dot{\boldsymbol{\theta}}] + \dot{\boldsymbol{\theta}} \cdot \mathbf{C}(\boldsymbol{\theta}) \cdot \dot{\boldsymbol{\theta}} + \mathbf{g}(\boldsymbol{\theta})$$
$$+ \mathbf{J}^T(\boldsymbol{\theta})(\mathbf{I} - \mathbf{S})\mathbf{K}_f \left(\frac{a}{s+a}\right)(\mathbf{f}_d - \mathbf{f})\}$$

$$\mathbf{x}_m^* = \ddot{\mathbf{x}}_d + \mathbf{K}_v(\dot{\mathbf{x}}_d - \dot{\mathbf{x}}) + \mathbf{K}_p(\mathbf{x}_d - \mathbf{x})$$

$\mathbf{F}(s)$ is the filter representing the torque servo loop at each joint and $\mathbf{K}_f$ is the wrist force sensor feedback gain matrix.

Joints 1 and 3 have inner torque loops operating at $500Hz$ and outer loops of both force and position operating at $100Hz$. The wrist force sensor feedback is included in order to increase the steady state force accuracy at the tip of the manipulator. In order to insure stability by creating a dominant pole, the force sensor signal is processed through a low pass filter, whose pole is at $1Hz$. A Barry Wright FS6-120A 6-axis wrist force/torque sensor was used.

Three types of inputs are used in the force control experiments:

1. square wave force in the x axis and constant position in the y axis,

2. sine wave force in the x axis and constant position in the y axis, and

3. constant position in the y axis and constant force in the x axis in presence of lateral movement of an eccentric cam (Figure 8.17).

In the results shown below, the measured force data from the wrist force sensor are plotted without any low-pass filtering of the signal.

The control parameters in the experiments are summarized below:

$$\mathbf{K}_f = \begin{bmatrix} 0.444 & 0 \\ 0 & 0.444 \end{bmatrix} \quad \mathbf{K}_p = \begin{bmatrix} 160 & 0 \\ 0 & 160 \end{bmatrix} \quad \mathbf{K}_v = \begin{bmatrix} 21 & 0 \\ 0 & 21 \end{bmatrix}$$

Kinematic Stability Issues in Force Control

The selection matrix $S = \text{diag}(0, 1)$. The force gain is set relatively low since the main dynamic performance should come from the joint torque control part.

9.3.2 Experimental Results

Square Wave Force Input

Under resolved acceleration, the x axis is force controlled with the square wave inputs of $10\,N$ and $15\,N$, and the y axis is position controlled at a stationary $y = -0.669\,m$. The step response is stable and fast (Figure 9.19), with a response time (delay plus rise time) of approximately $20\,ms$. The two axes are minimally coupled, as verified by the y position error. Except for some bias in the position error due to the deadzones in the motors and also probably due to the roundoff errors in the computations, there is very little change in y position as force steps are commanded along the x axis.

Sine Wave Force Input

Sine wave force response is used as a measure of bandwidth for the force controlled system. As in the step response experiments, the y axis is position controlled to be stationary at $y = -0.669\,m$, and the x axis is force controlled with the sine wave inputs of

$$f_{command_x}(N) = 12.5 + 2.5\sin(wt)$$

The force and position responses to the $1Hz$ sine force input are shown in Figure 9.20. Both the force and the position traces have some offset errors, but the controller performs well dynamically. The manipulator did not respond as well to higher frequency inputs, although the one-joint experiments presented in Chapter 8 showed good responses up to $20Hz$ inputs. This lower bandwidth is due to lower sampling frequencies ($100\,Hz$ outer force loop vs. $500\,Hz$) and less rigidity in the manipulator structure caused by the flexibilities at the joints connecting the links.

Following the Eccentric Cam

The manipulator is commanded to exert a constant force of $12.5\,N$ against the surface of the eccentric cam (Figure 8.17) in the x axis and stay stationary in the y axis at $y = -0.669\,m$. As the cam turns, it produces

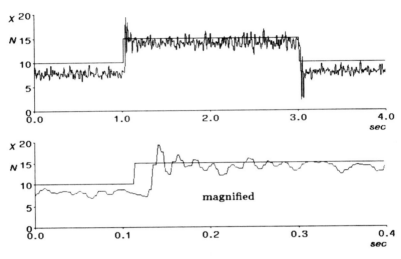

FORCE and TORQUE SENSING: X-AXIS STEP FORCE

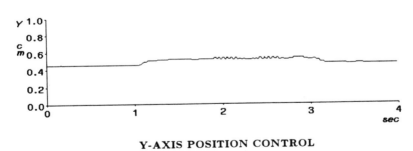

Y-AXIS POSITION CONTROL

Figure 9.19: Resolved acceleration: x axis=force steps, y axis=constant position.

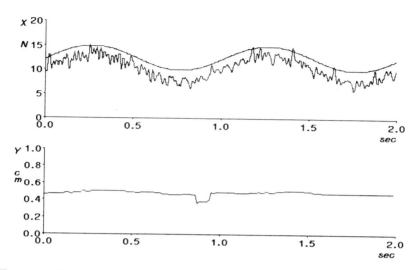

Figure 9.20: Resolved acceleration: x axis=$1\,Hz$ sine wave force, y axis=constant position.

a varying positional disturbance whose shape is approximately that of a sine wave.

The performance under these conditions is shown in Figure 9.21. The plot of the x axis position shows movements of the manipulator in the x direction, while maintaining a constant force against the moving cam. The position trace in the y axis shows that the resolved acceleration controller is able to maintain constant y position while following the cam in the x direction.

9.4 Discussion

In this chapter we have identified a new form of instability in force control, due to an inverse kinematic transformation in the feedback loop. This kinematic instability arises in hybrid control (Raibert and Craig, 1981), but not in stiffness control (Salisbury, 1980) or in modified resolved acceleration control (Luh, Walker, and Paul, 1980b; Shin and Lee, 1985).

The kinematic instabilities in hybrid control depend on the kinematic structure and configuration of a manipulator. With regard to kinematic

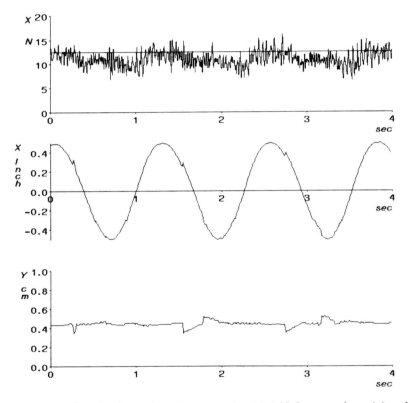

Figure 9.21: Resolved acceleration: x axis=12.5 N force and positional disturbance, y axis=constant position.

structure, we have shown that hybrid control is stable for a polar manipulator, but surprisingly it is unstable for a rotary manipulator. Raibert and Craig (1981) demonstrated hybrid control on two joints of the Stanford manipulator, equivalent to the polar manipulator in our analyses. Our analyses have confirmed their experimental stability results for hybrid control on this manipulator. With regard to kinematic configuration of a rotary manipulator, the kinematic instability appears only for certain nonsingular configurations. What happens is that the force-controlled joint drifts to a singular configuration, where the inverse Jacobian is undefined. This phenomenon does not depend on whether the manipulator is actually in contact with the environment or not. Altering feedback gains does not change this inherent unstable property, although the region of instability may become smaller. Furthermore, though the analysis was presented for a two degree-of-freedom rotary manipulator, the results generalize to rotary manipulators with more degrees of freedom.

The modified resolved acceleration control is stable because the inertia matrix is included and cancels the destabilizing effects of the inverse Jacobian. This latter control can be obtained through a trivial modification of resolved acceleration control (Luh, Walker, and Paul, 1980b) to include a force term (Shin and Lee, 1985). As such, resolved acceleration control is virtually identical to the operational space method (Khatib, 1983; Khatib and Burdick, 1986) and to impedance control (Hogan, 1985a-c). Hence the stability results hold for these equivalent controllers as well.

For resolved acceleration control, if the dynamic modeling is exact, the motion of the manipulator would be completely decoupled at the end effector in Cartesian coordinates. Then the response under the pure position control mode would be that of a unit mass along each Cartesian degree of freedom. For example, the resulting behavior in the x axis is

$$\ddot{x}_e + k_{vx}\dot{x}_e + k_{px}x_e = 0, \quad \text{where } x_e = x_d - x. \tag{9.30}$$

Even if the dynamic modeling is in error by 50%, our simulations show that resolved acceleration is still stable. Nevertheless, there is a limit to the tolerance to model inaccuracy. Replacing the inertia matrix with the identity matrix, for example, reduces resolved acceleration to hybrid control, which we now know is unstable. Thus a reasonably accurate dynamic model of the manipulator is as important in force control as it is in position control.

Stiffness control is stable because it uses the Jacobian transpose for coordinate transformations, rather than the inverse Jacobian. Though stiffness control and resolved acceleration control are both stable, in the simulations and experiments the best performance was obtained with the resolved acceleration controller since its inverse dynamics compensation decouples the motion of the tip of the manipulator in Cartesian directions.

Finally, we combined the results of our dynamic stability analyses of Chapter 8 and the kinematic stability analyses of this chapter to implement Cartesian force control during hard contact with a multi-joint manipulator. Resolved acceleration force control was implemented on two links of the DDArm, and the low-level force control law was comprised of open-loop joint torque control and low-pass filtered wrist force sensing. The inertial parameters of the robot had been identified in a previous experiment discussed in Chapter 5. The good results in step tracking, sinusoid tracking, and eccentric cam tracking emphasize that model-based control is important not only for position control, but for force control as well.

Chapter 10

Conclusion

This book has been about building models and using models to control a robot. Our hypothesis, born out by analysis and experiment, is that accurate modeling of the robot is important for control. We discovered that bad models can lead to instability in force control and learning, and to poor performance in position control.

For position control, the feedforward and computed torque controllers, both of which employ a dynamic model in the control structure, generally produced smaller trajectory errors in comparison to the independent-joint linear PD controllers, which do not employ any dynamic model.

In Cartesian-based force control, the major difference between resolved acceleration force control and hybrid position/force control is that the former uses a dynamic model of the robot. Hybrid position/force control can be derived from resolved acceleration force control by substituting the identity transformation for the dynamic model. A kinematic instability for rotary manipulators results because the identity transformation is not a good model for the robot.

Even though the Cartesian-based stiffness controller was found to be stable, it is not as accurate as resolved acceleration force control. Again, the stiffness controller does not contain a dynamic model of the robot. Since in Cartesian-based force control there are positions being generated as well as forces, it is entirely consistent that model-based controllers will also perform better for force control.

In trajectory learning, a learning operator that is the inverse of robot dynamics was seen to lead to rapid learning. Replacing this learning operator with an identity transformation, that is to say, ignoring the robot dynamics, can actually cause performance to degrade with practice.

Hence there was a major emphasis in this book on building accurate robot models. We sought model building methods that are not only accurate, but convenient and not demanding significant human involvement. Ideally, we look for methods that allow the robot to calibrate itself. In kinematic calibration, we employed a motion tracking system to automatically monitor endpoint position and orientation. Although the system used is not quite good enough to obtain accuracies below 1 mm, new motion tracking systems are much more accurate and reinforce the viability of this procedure. In inertial parameter estimation, our dynamic estimation procedure compared favorably to a CAD modeling approach.

These experiments also pointed out limitations in our ability to implement model-based control, primarily due to inaccuracy in joint torque control and in joint position sensing. As we pushed performance limits, flaws in our robot emerged rather than conceptual problems with the controllers. Unmodeled nonlinearities in joint 3 surfaced because of the light link structure, and manifested themselves as an equal level of performance of the PD controller and the feedforward controller.

10.1 Assessment of the DDArm

Our experiences with the DDArm have shown it to be an excellent device for laboratory experiments. Joint torques can be controlled fairly well, and rigid body dynamics were seen to approximate the robot system reasonably well. Until now, it would not have been possible to find a manipulator of comparable performance for research purposes. Thus without the DDArm the experiments reported here would not have been possible.

Our experiences also show that the DDArm is capable of high performance. Joint accelerations can top 6000 deg/sec^2, and this fast trajectory can be stably and accurately controlled with a model-based controller. Single trajectories are quite repeatable, and with the dynamically estimated link inertial parameters, learning occurs in just a few trials. Fast and stable force control has been demonstrated, such as a force step of 5 N executed in 20 msec.

At the same time, we encountered a number of difficulties with the DDArm. The motors are large, and together with the supporting structures make the arm quite heavy. The ISI motors are susceptible to over-

Conclusion

heating, and it was not possible to operate the arm continuously for a lengthy period. In particular, gravity as a continuous load could not be tolerated, even with the special configuration of shoulder roll motion.

Another problem is that the resolvers are on gears. Even though the arm joints themselves are direct drive, one still encounters with the resolvers the usual problems with gears. Although accuracy is nominally rated at 16 bits, we found that the actual accuracy was only 12-14 bits. This poses double trouble, since the resolvers are not only used for position control, but also for electrical commutation of the motor. Thus the ability to control joint torque is degraded, and in fact our joint torque control accuracy was no better than 5-10%.

The amplifiers to the motors also caused some difficulties. Dead zone was a problem around zero torque. At higher torques the input/output relation became nonlinear.

Since the initial foray in 1982 into direct drive technology with the ISI motors, there have been substantial improvements in direct drive technology. As of this writing, a number of advanced direct drive motors are now commercially available. Yokogawa Electric Corporation's Dynaserv Direct Drive Servo Actuator is rated at 3 N-m/kg motor torque to motor weight ratio, and its amplifier is said to allow joint torque control at a 1% accuracy level. Shin Meiwa Industry Ltd.'s High Performance Brushless DC Servo Motors are also a rare-earth magnet design, which have been used to build parallel drive, SCARA configuration direct drive arms for laser cutting applications (Asada and Youcef-Toumi, 1987).

A number of variable reluctance motors, pioneered at IBM (Pawletko and Chai, 1973), have an energy efficiency advantage over rare-earth magnet motor designs at present. The Motornetics Corporation's Megatorque Motor System, employed in the AdeptOne Robot and comprised of a double stator, is produced by the NSK Corporation. A single stator design is now produced by Superior Electric Company; their SLO-SYN Direct Drive Servo Positioning System permits torque control of 3% accuracy.

One can expect steady improvement in direct drive motor technology over the next few years, thereby overcoming some of the current performance limitations and leading to more practical designs. Magnetic materials have already been found that are several times stronger than those used in current motors. Improvements in amplifiers and electronics will continue. Of course, if superconductivity ever becomes practical at high temperatures, it would greatly advance direct direct motor technology.

A viable research strategy, therefore, may expect to update a direct drive arm design every few years to take advantage of improvements in motor technology. The modular design possible with direct drive technology makes such a construction relatively straightforward. While we have learned and accomplished a great deal with our current DDArm, it is now time to move on. We would like to see a new direct drive arm with greatly improved position and torque control. Even though the new motor designs allow joint torque control below 3% accuracy, we would like to see torque sensing on the motor axis. We would also like to see direct position sensing up to 20 bits of accuracy.

Because of the attractiveness of arm designs with electric motors, a direct drive arm will have a place in the laboratory, perhaps complemented by other arm designs. Hydraulic and pneumatic actuators are viable options for future arm designs. Alternative transmission systems to gears and to parallel linkages should be considered, such as tendons, currently used in multi-fingered robot hands (Jacobsen et al., 1986; Salisbury, 1985).

10.2 Further Issues

There are many issues surrounding the topics in this book that deserve further study. We regard the position control work as preliminary rather than definitive, given limitations in joint torque control and position sensing. The feedforward and computed torque controllers should be reevaluated against themselves and against independent-joint PD control with a better robot. The result that the computed torque controller was no more accurate than the feedforward controller needs to be further explored and explained.

The issue of sampling rate in computed torque control also merits reexamination. When we conducted the experiments, we were limited by the Motorola 68000 microprocessor system to a sampling rate of 133 Hz. Our computational architecture has been upgraded to a Motorola 68020 system, and in the future we should be able to elevate the sampling rate considerably. Thinking further into the future, special purpose real-time architectures based on digital signal processors look promising.

Force control certainly requires further study, especially the handling of collisions and the transition from unrestrained to restrained motions. Another issue is large positional deviations in the force-controlled direction, where the trajectory has to be controlled as well. While force control has been studied here in the context of arm movements, it is certainly an issue in hand movements as well (Hollerbach, Narasimhan, and Wood,

Conclusion

1986). The adequacy and completeness of our concepts will have to be tested for hand control. There also arises the issue of the variety of contact sensing: the role of endpoint sensing, tactile sensing, tendon force sensing, and proximal joint force sensing in force control. Another issue is combining finger and arm motions for coordinated force control.

In trajectory learning, the challenge will be to generalize learning for non-repeated trajectories. Learning can also be applied to tasks, such as visually-guided throwing (Atkeson et al., 1987).

In load inertial parameter estimation, one challenge will be to handle more complex loads than those characterized as rigid bodies. This would include loads that are flexible, articulated, or time varying (such as liquids in a container).

In modeling the manipulator itself, one would ideally like system identification methods that automatically derive structure. If a particular model such as a linear model does not characterize a motor well enough, for example, it would be nice to have robust techniques for extracting appropriate nonlinear models for the motor. This general system identification problem is a very difficult one, but restricted versions of this problem may be tractable and quite useful.

The research described in this book is only a first step towards true model-based control of robot manipulators. We look forward to improvements in robot designs and in the underlying hardware, and new ideas in system identification and control. The intelligent systems of tomorrow will depend on models of themselves and the world they deal with.

Appendices

Appendix 1: Integrating the Load Estimation Equations

In this appendix we will show how to integrate the load inertial parameter estimation equations (4.4) and (4.8). This integration is ultimately done in the force sensor frame so that we can integrate the measured load forces and torques directly. Throughout this derivation we will use the superscript notation ^{o} and ^{p} to indicate what coordinate frame a vector is expressed in. If there is no such superscript, the vector is expressed in the load frame.

The term $\ddot{\mathbf{p}}$ is integrated using equation 7.22 of (Symon, 1971):

$$\mathbf{R}\frac{d}{dt}(^o\dot{\mathbf{p}}) = \frac{d}{dt}(\mathbf{R} \cdot {^o\dot{\mathbf{p}}}) + \mathbf{R}(^o\boldsymbol{\omega} \times {^o\dot{\mathbf{p}}}) \tag{A.1}$$

were $\mathbf{R}$ is the rotation matrix representing the rotation from an inertial coordinate system to the force sensing coordinate system that continuously moves with the load. To get an integral form of this equation we simply integrate it:

$$\int_t^{t+T} \mathbf{R} \cdot {^o\ddot{\mathbf{p}}}\, d\tau = (\mathbf{R} \cdot {^o\dot{\mathbf{p}}}) \Big|_t^{t+T} + \int_t^{t+T} \mathbf{R}(^o\boldsymbol{\omega} \times {^o\dot{\mathbf{p}}})\, d\tau \tag{A.2}$$

and performing the indicated rotations leaves us with

$$\int_t^{t+T} {^p\ddot{\mathbf{p}}}\, d\tau = {^p\dot{\mathbf{p}}} \Big|_t^{t+T} + \int_t^{t+T} {^p\boldsymbol{\omega}} \times {^p\dot{\mathbf{p}}}\, d\tau \tag{A.3}$$

Similarly

$$\int_t^{t+T} {}^P\dot{\boldsymbol{\omega}}\, d\tau = {}^P\boldsymbol{\omega}\Big|_t^{t+T} + \int_t^{t+T} {}^P\boldsymbol{\omega} \times {}^P\boldsymbol{\omega}\, d\tau = {}^P\boldsymbol{\omega}\Big|_t^{t+T} \quad (A.4)$$

since $\boldsymbol{\omega} \times \boldsymbol{\omega} = 0$. We also remember that

$$\int_t^{t+T} {}^P\mathbf{g}\, d\tau = \left(\int_t^{t+T} \mathbf{R}\, d\tau\right) {}^o\mathbf{g} \quad (A.5)$$

Each of the matrices $[{}^P\dot{\boldsymbol{\omega}}\times]$ and $[\bullet\dot{\boldsymbol{\omega}}]$ can be integrated element by element to show that

$$\int_t^{t+T} [{}^P\dot{\boldsymbol{\omega}}\times]\, d\tau = \left[\left(\int_t^{t+T} {}^P\dot{\boldsymbol{\omega}}\, d\tau\right)\times\right] = \left[\left({}^P\boldsymbol{\omega}\Big|_t^{t+T}\right)\times\right] \quad (A.6)$$

$$\int_t^{t+T} [\bullet{}^P\dot{\boldsymbol{\omega}}]\, d\tau = \left[\bullet\left(\int_t^{t+T} {}^P\dot{\boldsymbol{\omega}}\, d\tau\right)\right] = \left[\bullet\left({}^P\boldsymbol{\omega}\Big|_t^{t+T}\right)\right] \quad (A.7)$$

The matrix $[(\mathbf{g} - \ddot{\mathbf{p}})\times]$ can be integrated element by element in the same way. Each matrix element of the terms $[{}^P\boldsymbol{\omega}\times][{}^P\boldsymbol{\omega}\times]$ and $[{}^P\boldsymbol{\omega}\times][\bullet{}^P\boldsymbol{\omega}]$ are numerically integrated by adding values at each time step.

We can express the resulting estimation equations in matrix form as:

$$\begin{bmatrix} \int_t^{t+T} \mathbf{f} \\ \int_t^{t+T} \mathbf{n} \end{bmatrix} = \left[\int_t^{t+T} \mathbf{A}\, d\tau\right] \begin{bmatrix} m \\ mc_x \\ mc_y \\ mc_z \\ I_{11} \\ I_{12} \\ I_{13} \\ I_{22} \\ I_{23} \\ I_{33} \end{bmatrix} \quad (A.8)$$

where the first row of $\left[\int_t^{t+T} \mathbf{A}\, d\tau\right]$ is

$$\left[\dot{\mathbf{p}}\Big|_t^{t+T} + \int_t^{t+T} \boldsymbol{\omega} \times \dot{\mathbf{p}}\, d\tau - \left(\int_t^{t+T} \mathbf{R}\, d\tau\right){}^o\mathbf{g} \quad \left[\left(\boldsymbol{\omega}\Big|_t^{t+T}\right)\times\right] + \int_t^{t+T} [\boldsymbol{\omega}\times][\boldsymbol{\omega}\times]\, d\tau \quad \mathbf{0}\right] \quad (A.9)$$

and the second row is

$$\begin{bmatrix} \mathbf{0} & \left[\left(-\dot{\mathbf{p}}\Big|_t^{t+T} - \int_t^{t+T} \boldsymbol{\omega} \times \dot{\mathbf{p}}\, d\tau + \left(\int_t^{t+T} \mathbf{R}\, d\tau\right){}^o\mathbf{g}\right) \times\right] \\ & \left[\bullet\left(\boldsymbol{\omega}\Big|_t^{t+T}\right)\right] + \int_t^{t+T} [\boldsymbol{\omega}\times][\bullet\boldsymbol{\omega}]\, d\tau \end{bmatrix} \quad (A.10)$$

Appendix 2: Closed Form Dynamics

The customized closed form equations of the dynamics of the DDArm are presented in this appendix. To simplify the equations, the following notation is used:

$$s_2 = \sin\theta_2, \quad s_3 = \sin\theta_3, \quad c_2 = \cos\theta_2, \quad c_3 = \cos\theta_3.$$

The closed form equations are:

$$\begin{aligned}
\tau_1 =\ & SP_4(\ddot{\theta}_1) + SP_5(2\dot{\theta}_1\dot{\theta}_2 s_2 c_2 - \ddot{\theta}_1 c_2^2) \\
& + I_{xy_2}(2\ddot{\theta}_1 s_2 c_2 - 2\dot{\theta}_1\dot{\theta}_2 + 4\dot{\theta}_1\dot{\theta}_2 c_2^2) \\
& + I_{xz_2}(\ddot{\theta}_2 s_2 + \dot{v}_2^2 c_2) + SP_1(\ddot{\theta}_2 c_2 - \dot{v}_2^2 s_2) \\
& + m_3 c_{x_3}(l_2)(2\ddot{\theta}_1 s_3 + 2\dot{\theta}_2\dot{\theta}_3 s_2 s_3 - \ddot{\theta}_3 s_3 c_2 + 2\dot{\theta}_1\dot{\theta}_3 c_3 - \ddot{\theta}_2 s_2 c_3 \\
& \quad - \dot{v}_2^2 c_2 c_3 - \dot{v}_3^2 c_2 c_3) \\
& + m_3 c_{y_3}(l_2)(\ddot{\theta}_2 s_2 s_3 - 2\dot{\theta}_1\dot{\theta}_3 s_3 + \dot{v}_2^2 s_3 c_2 + \dot{v}_3^2 s_3 c_2 + 2\ddot{\theta}_1 c_3 \\
& \quad + 2\dot{\theta}_2\dot{\theta}_3 s_2 c_3 - \ddot{\theta}_3 c_2 c_3) \\
& + SP_2(\ddot{\theta}_1 c_2^2 - \ddot{\theta}_1 - \dot{\theta}_2\dot{\theta}_3 s_2 - 2\dot{\theta}_1\dot{\theta}_2 s_2 c_2 - 2\dot{\theta}_1\dot{\theta}_3 s_3 c_3 + \ddot{\theta}_2 s_2 s_3 c_3 \\
& \quad + \dot{v}_2^2 s_3 c_2 c_3 + 2\dot{\theta}_1\dot{\theta}_3 s_3 c_3 c_2^2 + \ddot{\theta}_1 c_3^2 + 2\dot{\theta}_2\dot{\theta}_3 s_2 c_3^2 + 2\dot{\theta}_1\dot{\theta}_2 s_2 c_2 c_3^2 \\
& \quad - \ddot{\theta}_1 c_2^2 c_3^2) \\
& + I_{xy_3}(2\ddot{\theta}_1\dot{\theta}_3 - \ddot{\theta}_2 s_2 - \dot{v}_2^2 c_2 - 2\dot{\theta}_1\dot{\theta}_3 c_2^2 - 2\ddot{\theta}_1 s_3 c_3 - 4\dot{\theta}_2\dot{\theta}_3 s_2 s_3 c_3 \\
& \quad - 4\dot{\theta}_1\dot{\theta}_2 s_2 s_3 c_2 c_3 + 2\ddot{\theta}_1 s_3 c_2^2 c_3 - 4\dot{\theta}_1\dot{\theta}_3 c_3^2 + 2\ddot{\theta}_2 s_2 c_3^2 + 2\dot{\theta}_2\dot{\theta}_2 c_2 c_3^2 \\
& \quad + 4\dot{\theta}_1\dot{\theta}_3 c_2 c_2 c_3^2) \\
& + I_{xz_3}(\ddot{\theta}_2\dot{\theta}_2 s_2 s_3 - \dot{v}_3^2 s_2 s_3 - \ddot{\theta}_2 s_3 c_2 + 2\dot{\theta}_1\dot{\theta}_3 s_2 s_3 c_2 + 2\dot{\theta}_1\dot{\theta}_2 c_3 \\
& \quad + \ddot{\theta}_3 s_2 c_3 - 2\dot{\theta}_1 s_2 c_2 c_3 - 4\dot{\theta}_1\dot{\theta}_2 c_2^2 c_3) \\
& + I_{yz_3}(2\ddot{\theta}_1 s_2 s_3 c_3 - 2\dot{\theta}_1\dot{\theta}_2 s_3 - \ddot{\theta}_3 s_2 s_3 + 4\dot{\theta}_1\dot{\theta}_2 s_3 c_2^2 + \dot{v}_2^2 s_2 c_3 \\
& \quad - \dot{v}_3^2 s_2 c_3 - \ddot{\theta}_2 c_2 c_3 + 2\dot{\theta}_1\dot{\theta}_3 s_2 c_2 c_3) \\
& + I_{zz_3}(\ddot{\theta}_2\dot{\theta}_3 s_2 - \ddot{\theta}_3 c_2 - 2\dot{\theta}_1\dot{\theta}_2 s_2 c_2 + \ddot{\theta}_1 c_2^2)
\end{aligned}$$

Closed Form Dynamics

$$\begin{aligned}
\tau_2 = & \; m_2 c_{x_2}(gc_2) + \mathrm{SP}_6(gs_2) - \mathrm{SP}_5(v_1^2 s_2 c_2) \\
& + I_{xy_2}(v_1^2 - 2v_1^2 c_2^2) + I_{xz_2}(\ddot{\theta}_1 s_2) \\
& + \mathrm{SP}_1(\ddot{\theta}_1 c_2) + \mathrm{SP}_3(\ddot{\theta}_2) + m_3 c_{x_3}(gc_2 c_3 - l_2 \ddot{\theta}_1 s_2 c_3) \\
& + m_3 c_{y_3}(l_2 \ddot{\theta}_1 s_2 s_3 - gs_3 c_2) \\
& + \mathrm{SP}_2(\dot{\theta}_1 \dot{\theta}_1 s_2 c_2 - \dot{\theta}_1 \dot{\theta}_3 s_2 + 2\dot{\theta}_2 \dot{\theta}_3 s_3 c_3 + \ddot{\theta}_1 s_2 s_3 c_3 - \ddot{\theta}_2 c_3^2 \\
& + 2\dot{\theta}_1 \dot{\theta}_3 s_2 c_3^2 - \dot{\theta}_1 \dot{\theta}_1 s_2 c_2 c_3^2) \\
& + I_{xy_3}(2\ddot{\theta}_2 s_3 c_3 - 2\dot{\theta}_2 \dot{\theta}_3 - \ddot{\theta}_1 s_2 - 4\dot{\theta}_1 \dot{\theta}_3 s_2 s_3 c_3 + 2v_1^2 s_2 s_3 c_2 c_3 \\
& + 4\dot{\theta}_2 \dot{\theta}_3 c_3^2 + 2\ddot{\theta}_1 s_2 c_3^2) \\
& + I_{xz_3}(\ddot{\theta}_3 s_3 - \ddot{\theta}_1 s_3 c_2 - v_1^2 c_3 + v_3^2 c_3 - 2\dot{\theta}_1 \dot{\theta}_3 c_2 c_3 + 2v_1^2 c_2^2 c_3) \\
& + I_{yz_3}(v_1^2 s_3 - v_3^2 s_3 + 2\dot{\theta}_1 \dot{\theta}_3 s_3 c_2 - 2v_1^2 s_3 c_2^2 + \ddot{\theta}_3 c_3 - \ddot{\theta}_1 c_2 c_3) \\
& + I_{zz_3}(v_1^2 s_2 c_2 - \dot{\theta}_1 \dot{\theta}_3 s_2)
\end{aligned}$$

$$\begin{aligned}
\tau_3 = & \; m_3 c_{x_3}(-gs_2 s_3 - l_2 \ddot{\theta}_1 s_3 c_2 - l_2 v_1^2 c_3) \\
& + m_3 c_{y_3}(l_2 v_1^2 s_3 - l_2 \ddot{\theta}_1 c_2 c_3 - gs_2 c_3) \\
& + \mathrm{SP}_2(\dot{\theta}_1 \dot{\theta}_2 s_2 + v_1^2 s_3 c_3 - v_2^2 s_3 c_3 - v_1^2 s_3 c_3 c_2^2 - 2\dot{\theta}_1 \dot{\theta}_2 s_2 c_3^2) \\
& + I_{xy_3}(v_2^2 - v_1^2 + v_1^2 c_2^2 + 4\dot{\theta}_1 \dot{\theta}_2 s_2 s_3 c_3 + 2v_1^2 c_3^2 - 2v_2^2 c_3^2 - 2v_1^2 c_2^2 c_3^2) \\
& + I_{xz_3}(\ddot{\theta}_2 s_3 - v_1^2 s_2 s_3 c_2 + \ddot{\theta}_1 s_2 c_3 + 2\dot{\theta}_1 \dot{\theta}_2 c_2 c_3) \\
& + I_{yz_3}(\ddot{\theta}_2 c_3 - \ddot{\theta}_1 s_2 s_3 - 2\dot{\theta}_1 \dot{\theta}_2 s_3 c_2 - v_1^2 s_2 c_2 c_3) \\
& + I_{zz_3}(\ddot{\theta}_3 + \dot{\theta}_1 \dot{\theta}_2 s_2 - \ddot{\theta}_1 c_2)
\end{aligned}$$

In these equations, there are 15 reduced inertial parameters: $m_2 c_{x_2}$, I_{xy_2}, I_{xz_2}, $m_3 c_{x_3}$, $m_3 c_{y_3}$, I_{xy_3}, I_{xz_3}, I_{yz_3}, I_{zz_3}, SP_1, SP_2, SP_3, SP_4, SP_5, SP_6. The SP_i variables are abbreviations for the following linear combinations:

$$\begin{aligned}
\mathrm{SP}_1 &= m_3 c_{z_3} l_2 + I_{yz_2} \\
\mathrm{SP}_2 &= I_{xx_3} - I_{yy_3} \\
\mathrm{SP}_3 &= I_{zz_2} + I_{xx_3} \\
\mathrm{SP}_4 &= I_{zz_1} + I_{xx_2} + I_{xx_3} + m_3 l_2^2 \\
\mathrm{SP}_5 &= I_{xx_2} + I_{xx_3} - I_{yy_2} \\
\mathrm{SP}_6 &= m_3 c_{z_3} - m_2 c_{y_2}
\end{aligned}$$

Eleven inertial parameters do not appear in these equations at all, and are completely unidentifiable: m_1, $m_1 c_{x_1}$, $m_1 c_{y_1}$, $m_1 c_{z_1}$, I_{xx_1}, I_{xy_1}, I_{xz_1}, I_{yy_1}, I_{yz_1}, m_2, $m_2 c_{z_2}$.

Appendix 3: Stability Robustness

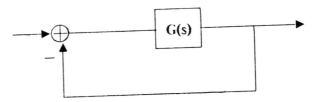

Figure A.1: Nominal SISO system.

This appendix is a summary of the discussion in Lehtomaki (1981) on the stability robustness using the Nyquist criterion for a single input single output (SISO) system,

Consider a SISO system described in Figure A.1 with nominal loop transfer function $G(s)$. If we have a modelling error, the actual loop transfer function is represented as the perturbed $\tilde{G}(s)$. The stability robustness is determined by the distance that the Nyquist plot of $\tilde{G}(s)$ avoids the (-1,0) point in the complex plane (Figure A.2). The situation in Figure A.2 shows that if the nominal closed-loop system with $G(s)$ were stable, then the perturbed system with $\tilde{G}(s)$ would also be stable since the number of encirclements of the (-1, 0) point has not changed.

For any ω, the distance between the (-1, 0) point and $G(j\omega)$ is given by

$$d(w) = |1 + G(j\omega)|, \qquad (A.11)$$

and the distance between $\tilde{G}(j\omega)$ and $G(j\omega)$ is

$$p(w) = |\tilde{G}(j\omega) - G(j\omega)|. \qquad (A.12)$$

Then from the Nyquist plot it is clear that the perturbed closed-loop system is stable if

$$|1 + G(j\omega)| > |\tilde{G}(j\omega) - G(j\omega)|. \qquad (A.13)$$

If we define the modelling error as in Figure 8.2, then

$$\tilde{G}(s) = (1 + E(s))G(s)$$

and

$$E(s) = \frac{\tilde{G}(s) - G(s)}{G(s)}.$$

Stability Robustness

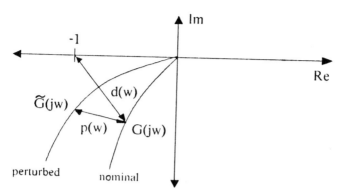

Figure A.2: Nyquist plot of the nominal and the perturbed (actual) system.

The relation for stability robustness is obtained for this error model by dividing (A.13) by $G(j\omega)$:

$$\left|\frac{1+G(j\omega)}{G(j\omega)}\right| = |1+G^{-1}(j\omega)| > \left|\frac{\tilde{G}(j\omega)-G(j\omega)}{G(j\omega)}\right| = |E(j\omega)|. \quad (A.14)$$

Appendix 4: Operational Space and Resolved Acceleration

In the reports by Khatib (Khatib, 1983; Khatib and Burdick, 1986), the dynamics of a manipulator in operational space or end effector cartesian space are described by

$$\Lambda(\mathbf{x})\ddot{\mathbf{x}} + \mu(\mathbf{x}, \dot{\mathbf{x}}) + \mathbf{p}(\mathbf{x}) = \mathbf{f}. \qquad [1]^1$$

In joint coordinate system, the same dynamics are described by

$$\mathbf{H}(\mathbf{q})\ddot{\mathbf{q}} + \dot{\mathbf{q}} \cdot \mathbf{C}(\mathbf{q}) \cdot \dot{\mathbf{q}} + \mathbf{g}(\mathbf{q}) = \tau. \qquad (A.15)$$

$\mathbf{H}(\mathbf{q})$ and $\Lambda(\mathbf{x})$ are related by

$$\mathbf{H}(\mathbf{q}) = \mathbf{J}^T(\mathbf{q})\Lambda(\mathbf{x})\mathbf{J}(\mathbf{q}) \qquad [3]$$

or

$$\mathbf{J}^{-T}(\mathbf{q})\mathbf{H}(\mathbf{q})\mathbf{J}^{-1}(\mathbf{q}) = \Lambda(\mathbf{x}) = \Lambda(\mathbf{q}). \qquad (A.16)$$

Also, the operational space force vector $\mathbf{f}$ and the joint torque vector τ are related by

$$\tau = \mathbf{J}^T(\mathbf{q})\mathbf{f}. \qquad [4]$$

For a decoupled end effector motion commanded by $\mathbf{f}_m^*$,

$$\begin{aligned}\tau &= \mathbf{J}^T(\mathbf{q})\Lambda(\mathbf{q})\mathbf{f}_m^* + \mathbf{J}^T\mu(\mathbf{x},\dot{\mathbf{x}}) + \mathbf{J}^T\mathbf{p}(\mathbf{x}) \\ &= \mathbf{J}^T(\mathbf{q})\Lambda(\mathbf{q})\mathbf{f}_m^* + \dot{\mathbf{q}} \cdot \tilde{\mathbf{C}}(\mathbf{q}) \cdot \dot{\mathbf{q}} + \mathbf{g}(\mathbf{q}).\end{aligned} \qquad [7]$$

where

$$\dot{\mathbf{q}} \cdot \tilde{\mathbf{C}}(\mathbf{q}) \cdot \dot{\mathbf{q}} = \dot{\mathbf{q}} \cdot \mathbf{C}(\mathbf{q}) \cdot \dot{\mathbf{q}} - \mathbf{J}^T\Lambda(\mathbf{q})\mathbf{h}(\mathbf{q},\dot{\mathbf{q}}), \qquad [8]$$

$$\mathbf{h}(\mathbf{q},\dot{\mathbf{q}}) = \dot{\mathbf{J}}(\mathbf{q})\dot{\mathbf{q}}. \qquad [9]$$

Typically, for position or trajectory control, a linear second order behavior is commanded so that

$$\mathbf{f}_m^* = \ddot{\mathbf{x}}_d + \mathbf{K}_v(\dot{\mathbf{x}}_d - \dot{\mathbf{x}}) + \mathbf{K}_p(\mathbf{x}_d - \mathbf{x}) \qquad (A.17)$$

Then, for a hybrid position/force control, the joint command vector with the operational space method is

$$\tau = \mathbf{J}^T(\mathbf{q})[\Lambda(\mathbf{q})\mathbf{S}\mathbf{f}_m^* + (\mathbf{I} - \mathbf{S})\mathbf{f}_a^*] + \dot{\mathbf{q}} \cdot \tilde{\mathbf{C}}(\mathbf{q}) \cdot \dot{\mathbf{q}} + \mathbf{g}(\mathbf{q}) \qquad [18]$$

---

[1] In this appendix, the equation numbers in brackets refer to the equation numbers in (Khatib and Burdick, 1986)

where $\mathbf{f}_a^*$ is the command vector for force control.

Substituting (A.16) and [8] into [18],

$$\begin{aligned}\boldsymbol{\tau} = {} & \mathbf{J}^T(\mathbf{q})\mathbf{J}^{-T}(\mathbf{q})\mathbf{H}(\mathbf{q})\mathbf{J}^{-1}(\mathbf{q})\mathbf{S}\mathbf{f}_m^* + \dot{\mathbf{q}} \cdot \mathbf{C}(\mathbf{q}) \cdot \dot{\mathbf{q}} + \mathbf{g}(\mathbf{q}) \\ & -\mathbf{J}^T\mathbf{J}^{-T}(\mathbf{q})\mathbf{H}(\mathbf{q})\mathbf{J}^{-1}(\mathbf{q})\mathbf{h}(\mathbf{q},\dot{\mathbf{q}}) + \mathbf{J}^T(\mathbf{q})(\mathbf{I}-\mathbf{S})\mathbf{f}_a^*.\end{aligned} \quad (A.18)$$

Simplifying the above equation,

$$\boldsymbol{\tau} = \mathbf{H}(\mathbf{q})\mathbf{J}^{-1}(\mathbf{q})[\mathbf{S}\mathbf{f}_m^* - \mathbf{h}(\mathbf{q},\dot{\mathbf{q}})] + \dot{\mathbf{q}}\cdot\mathbf{C}(\mathbf{q})\cdot\dot{\mathbf{q}} + \mathbf{g}(\mathbf{q}) + \mathbf{J}^T(\mathbf{q})(\mathbf{I}-\mathbf{S})\mathbf{f}_a^*. \quad (A.19)$$

This equation is identical to the equation (9.4) for the modified resolved acceleration controller presented in Chapter 9.

References

An, C. H. 1986. Trajectory and Force Control of a Direct Drive Arm. TR 912. Cambridge, MA: MIT Artificial Intelligence Laboratory.

An, C. H. 1986 (Sept.). Trajectory and Force Control of a Direct Drive Arm. Ph.D. thesis, MIT, Department of Electrical Engineering and Computer Science.

An, C. H., Atkeson, C. G., Griffiths, J. D., and Hollerbach, J. M. 1986 (September 9-12). Experimental evaluation of feedforward and computed torque control. *6th CISM-IFToMM Symposium on Theory and Practice of Robots and Manipulators*. Cracow, Poland: pp. 434-441.

An, C. H., Atkeson, C. G., Griffiths, J. D., and Hollerbach, J. M. 1987 (Mar. 31–Apr. 2). Experimental evaluation of feedforward and computed torque control. *Proc. IEEE Int. Conf. Robotics and Automation.* Raleigh, N.C.: pp. 165-168.

An, C. H., Atkeson, C. G., and Hollerbach, J. M. 1985 (Dec. 11-13). Estimation of inertial parameters of rigid body links of manipulators. *Proc. 24th IEEE Conf. Decision and Control.* Fort Lauderdale: pp. 990-995.

An, C. H., Atkeson, C. G., and Hollerbach, J. M. 1986 (April 7-10). Experimental determination of the effect of feedforward control on trajectory tracking errors. *Proc. IEEE Conf. on Robotics and Automation.* San Francisco: pp. 55-60.

An, C. H., and Hollerbach, J. M. 1987. Dynamic stability issues in force control of manipulators. *Proc. IEEE Int. Conf. Robotics & Auto.*. Raleigh: pp. 890-896.

An, C. H., and Hollerbach, J. M. 1987 (March 30 - April 3). Kinematic stability issues in force control of manipulators. *Proc.IEEE Int. Conf. Robotics and Automation.* Raleigh: pp. 897-903.

Antonsson, E. K. 1982. A three-dimensional kinematic acquisition and intersegmental dynamic analysis system for human motion. Ph. D. thesis, Massachusetts Institute of Technology, Mechanical Engineering.

Arimoto, S. 1985 (May). Mathematical Theory of Learning With Applications to Robot Control. *Proc. of 4th Yale Workshop on Applications of Adaptive Systems Theory.* Center for Systems Science, Yale University: pp. 215-220.

Arimoto, S., Kawamura, S., and Miyazaki, F. 1984a. Bettering Operation of Robots by Learning. *J. of Robotic Systems.* 1(2): 123-140.

Arimoto, S., Kawamura, S., and Miyazaki, F. 1984b (August). Can Mechanical Robots Learn by Themselves?. *Proc. of 2nd Inter. Symp. Robotics Research.* Kyoto, Japan: pp. 127-134.

Arimoto, S., Kawamura, S., and Miyazaki, F. 1984c (Dec). Bettering Operation of Dynamic Systems By Learning: A New Control Theory For Servomechanisms of Mechatronics Systems. *Proc. 23rd IEEE CDC.* Las Vegas, Nevada: pp. 1064-1069.

Arimoto, S., Kawamura, S., and Miyazaki, F., and Tamaki, S. 1985 (Dec. 11-13). Learning Control Theory for Dynamical Systems. *Proc. 24th Conf. on Decision and Control.* Fort Lauderdale, Florida: pp. 1375-1380.

Armstrong, B. 1987 (Mar 31 - April 3). On finding 'exciting' trajectories for identification experiments involving systems with non-linear dynamics. *Proc. IEEE Int. Conf. Robotics and Automation.* Raleigh: pp. 1131-1139.

Armstrong, B., Khatib, O. and Burdick, J. 1986 (April 7-10). The explicit dynamic model and inertial parameters of the PUMA 560 arm. *Proc. IEEE Int. Conf. Robotics and Automation.* San Francisco: pp. 510-518.

References

Asada, H., and Kanade, T. 1981 (April 28). Design of Direct-Drive Mechanical Arms. CMU-RI-TR-81-1. Carnegie-Mellon Univ., Robotics Institute.

Asada, H., Kanade, K., and Takeyama, I. 1983 (Sept). Control of a direct-drive arm. *Trans. of ASME.* 105: 136-142.

Asada, H., and Ro, I. H. 1985. A linkage design for direct-drive robot arms. *J. Mechanisms, Transmissions, and Automation in Design.* 107: 536-540.

Asada, H., and Youcef-Toumi, K. 1984. Analysis and design of a direct-drive arm with a five-bar-link parallel drive mechanism. *ASME J. Dynamic Systems, Meas., Control.* 106: 225-230.

Asada, H., Youcef-Toumi, K., and Lim, S. K. 1984 (Dec. 12-14). Joint torque measurement of a direct-drive arm. *Proc. 23rd IEEE Conf. Decision and Control.* Las Vegas: pp. 1332-1337.

Asada, H., and Youcef-Toumi, K. 1987. *Direct Drive Robots: Theory and Practice.* Cambridge, MA: MIT Press.

Åström, K. J. and Wittenmark, B. 1984. *Computer Controlled Systems: Theory and Design.* Englewood Cliffs, N.J.: Prentice-Hall, Inc..

Atkeson, C. G. 1986 (Sept.). Roles of Knowledge in Motor Learning. Ph.D. thesis, MIT, Dept. of Brain and Cognitive Sciences.

Atkeson, C. G., Aboaf, E. W., McIntyre, J., and Reinkensmeyer, D. J. 1987 (August). Model-Based Robot Learning. *Proc. of 4th Inter. Symp. Robotics Research.* Santa Cruz, CA:.

Atkeson, C. G., An, C. H., and Hollerbach, J. M. 1985a (October 7-11). Estimation of inertial parameters of manipulator loads and links. *Preprints of 3rd Int. Symp. Robotics Research.* Gouvieux (Chantilly), France: pp. 32-39.

Atkeson, C. G., An, C. H., and Hollerbach, J. M. 1985b (Dec. 11-13). Rigid body load identification for manipulators. *Proc. 24th Conf. Decision and Control.* Fort Lauderdale: pp. 996-1002.

Atkeson, C. G., An, C. H., and Hollerbach, J. M. 1986. Estimation of inertial parameters of manipulator loads and links. *Int. Journal of Robotics Research.* 5(3): 101-119.

Atkeson, C. G., and McIntyre, J. A. 1986 (July 1-3). Applications of Adaptive Feedforward Control in Robotics. *Proceedings, 2nd IFAC Workshop on Adaptive Systems in Control and Signal Processing.* Lund, Sweden:.

Atkeson, C. G., and McIntyre, J. A. 1986 (April 7-10). Robot trajectory learning through practice. *Proc. IEEE Int. Conf. Robotics and Automation.* San Francisco: pp. 1737-1742.

Baker, D. R., and Wampler, C. W. 1987 (March 31-April 3). Some facts concerning the inverse kinematics of redundant manipulators. *Proc. IEEE Int. Conf. Robotics and Automation.* Raleigh, NC: pp. 604-609.

Bejczy, A. K., Tarn, T. J., and Chen, Y. L. 1985 (Mar. 25-28). Robot arm dynamic control by computer. *Proc. IEEE Conf. Robotics and Automation.* St. Louis: pp. 960-970.

Brady, M., Hollerbach, J. M., Johnson, T. L., Lozano-Perez, T., Mason, M., editors 1982. *Robot Motion: Planning and Control.* Cambridge, MA: MIT Press.

Caine, M. E. 1985 (June). Chamferless Assembly of Rectangular Parts in Two and Three Dimensions. SM thesis, MIT, Mechanical Engineering Department.

Casalino, G., and Gambardella, L. 1986 (April 7-10). Learning of Movements in Robotic Manipulators. *Proc. 1986 IEEE International Conference on Robotics and Automation.* San Francisco, CA: pp. 572-578.

Chao, L. M., and Yang, J. C. S. 1986 (April 20-24). Development and implementation of a kinematic parameter identification technique to improve the positioning accuracy of robots. *Robots 10 Conference Proceedings.* Chicago: pp. 11-69 – 11-81.

Chen, Y. 1987 (March 31-April 3). Frequency response of discrete-time robot systems – limitations of PD controllers and improvements by lag-lead compensation. *Proc. IEEE Int. Conf. Robotics and Automation.* Raleigh, NC: pp. 464-472.

Chen, J., and Chao, L. M. 1986 (April 7-10). Positioning error analysis for robot manipulators with all rotary joints. *Proc. IEEE Int. Conf. Robotics and Automation.* San Francisco: pp. 1011-1016.

Coiffet, P. 1983. *Robot Technology: Interaction With The Environment. Vol. 2.* Englewood Cliffs, N.J.: Prentice-Hall.

Craig, J. J. 1984 (June 6-8). Adaptive Control of Manipulators Through Repeated Trials. *Proc. American Control Conference.* San Diego: pp. 1566-1574.

Curran, R., and Mayer, G. 1985 (June 19-21). The architecture of the AdeptOne direct-drive robot. *Proc. American Control Conf..* Boston: pp. 716-721.

Dagalakis, N. G., and Myers, D. R. 1985. Adjustment of robot joint gear backlash. *Int. J. Robotics Research.* 4(2): 65-79.

Dainis, A., and Juberts, M. 1985 (March 25-28). Accurate remote measurement of robot trajectory motion. *IEEE Int. Conf. Robotics and Automation.* St. Louis: pp. 92-99.

Denavit, J., and Hartenberg, R. S. 1955. A kinematic notation for lower pair mechanisms based on matrices. *J. Applied Mechanics.* 22: 215-221.

De Schutter, J. 1986 (February). Compliant Robot Motion: Task Formulation and Control. Ph. D. thesis, Katholieke Universiteit Leuven, Departement Werktuigkunde.

El-Zorkany, H. I., Liscano, R., and Tondu, B. 1985. A sensor-based approach for robot programming. *Int. Conf. Intelligent Robots and Computer Vision, SPIE Proc. Vol. 579.* Cambridge, Mass.: pp. 289-297.

Eppinger, S. D., and Seering, W. P. 1986 (April 7-10). On dynamic models of robot force control. *Proc. IEEE Int. Conf. Robotics and Automation.* San Francisco: pp. 29-34.

Eppinger, S. D., and Seering, W. P. 1987 (Mar. 31–Apr. 2). Understanding bandwidth limitations in robot force control. *Proc. IEEE Int. Conf. Robotics and Automation.* Raleigh, N.C.: pp. 904-909.

Eveleigh, V. W. 1972. *Introduction to Control Systems Design.* New York:

Mcgraw-Hill Book Company.

Foulloy, L. P., and Kelley, R. B. 1984 (March 13-15). Improving the precision of a robot. *Proc. IEEE Int. Conf. Robotics.* Atlanta: pp. 62-67.

Franklin, G. F. and Powel, J. D. 1980. *Digital Control of Dynamic Systems.* Reading, MA: Addison-Wesley.

Furuta, K., and Yamakita, M. 1986a (April 7-10). Iterative Generation of Optimal Input of a Manipulator. *Proc. 1986 IEEE International Conference on Robotics and Automation.* San Francisco, CA: pp. 579-584.

Gilbert, E. G., and Ha, I. J. 1984. An approach to nonlinear feedback control with applications to robotics. *IEEE Trans. Systems, Man, Cybern.* SMC-14: 879-884.

Gilby, J. H., and Parker, G. A. 1982. Laser tracking system to measure robot arm performance. *Sensor Review.* 2(4): 180-184.

Golub, G. H., and Van Loan, C. F. 1983. *Matrix computations.* John Hopkins University Press.

Good, M. C., Sweet, L. M., and Strobel, K. L. 1985. Dynamic models for control system design of integrated robot and drive systems. *ASME J. Dynamic Systems, Meas., Control.* 107: 53-59.

Goor, R. M. 1985a (June 19-21). A new approach to robot control. *Proc. American Control Conf.* Boston: pp. 385-389.

Goor, R. M. 1985b (Nov. 17-22). A new approach to minimum time robot control. *ASME Winter Annual Meeting.* Miami Beach, Florida: pp. 1-12.

Griffiths, J. D. 1986. Experimental Evaluation of Computed Torque Control. B.S. thesis, MIT, Mechanical Eng..

Hara, S., Omata, T., and Nakano, M. 1985 (Dec. 11-13). Synthesis of Repetitive Control Systems and its Application. *Proc. 24th Conf. on Decision and Control.* Fort Lauderdale, Florida: pp. 1387-1392.

Harokopos, E. G. 1986a (April 7-10). Optimal Learning Control of Mechanical Manipulators in Repetitive Motions. *Proc. 1986 IEEE International Conference on Robotics and Automation.* San Francisco, CA: pp. 396-401.

Harokopos, E. G. 1986b (April 20-24). Learning and Optimal Control of Industrial Robots in Repetitive Motions. *Proc. Robots 10.* Chicago, Illinois: pp. 4-27 - 4-46.

Hayati, S. A. 1983 (Dec. 14-16). Robot arm geometric link parameter estimation. *Proc. 22nd IEEE Conf. Decision and Control.* San Antonio: pp. 1477-1483.

Hayati, S. A., and Mirmirani, M. 1985. Improving the absolute positioning accuracy of robot manipulators. *J. Robotic Systems.* 2: 397-413.

Hayati, S. A., and Roston, G. P. 1986. Inverse kinematic solution for near-simple robots and its application to robot calibration. *Recent Trends in Robotics: Modeling, Control, and Education.*, ed. M. Jamshidi, J.Y.S. Luh, and M. Shahinpoor. Elsevier Science Publ. Co., pp. 41-50.

Hogan, N. 1985a (March). Impedance control: An approach to manipulation: Part I - Theory. *ASME J. of Dynamic Systems, Measurement, and Control.* 107: 1-7.

Hogan, N. 1985b (March). Impedance control: An approach to manipulation: Part II - Implementation. *ASME J. of Dynamic Systems, Measurement, and Control.* 107: 8-16.

Hogan, N. 1985c (March). Impedance control: An approach to manipulation: Part III - Applications. *ASME J. of Dynamic Systems, Measurement, and Control.* 107: 17-24.

Hollerbach, J. M., Narasimhan, S., and Wood, J. E. 1986 (April 7-10). Finger force computation without the grip Jacobian. *Proc. IEEE Int. Conf. Robotics and Automation.* San Francisco: pp. 871-875.

Hollerbach, J. M., and Sahar, G. 1983. Wrist-partitioned inverse kinematic accelerations and manipulator dynamics. *Int. J. Robotics Research.* 2(4): 61-76.

Hollerbach, J. M., and Suh, K. C. 1987. Redundancy resolution of manipulators through torque optimization. *IEEE J. Robotics and Automation.* 3(4): in press.

Hsu, P., Bodson, M., Sastry, S., and Paden, B. 1987 (Mar 31 - April 3). Adaptive identification and control for manipulators without using joint accelerations. *Proc. IEEE Int. Conf. Robotics and Automation.* Raleigh: pp. 1201-1215.

Isaacson, E., and Keller, H. B. 1966. *Analysis of Numerical Methods.* New York: John Wiley and Sons.

Ishii, M., Sakane, S., Mikami, Y., and Kakikura, M. 1987. A 3-D sensor system for teaching robot paths and environments. *Int. J. Robotics Research.* 6(2): 45-59.

Ish-Shalom, J., and Manzer, D. 1985. IBM Research Report. RC 11084. IBM.

Jacobsen, S. C., Iversen, E. K., Knutti, D. F., Johnson, R. T., and Biggers, K. B. 1986 (April 7-10). Design of the Utah/MIT dexterous hand. *Proc. IEEE Conf. on Robotics and Automation.* San Francisco: pp. 1520-1532.

Judd, R.P., and Knasinski, A.B. 1987 (March 31-April 3). A technique to calibrate industrial robots with experimental verification. *Proc. IEEE Int. Conf. Robotics and Automation.* Raleigh, NC: pp. 351-357.

Kanade, T., Khosla, P. K., and Tanaka, N. 1984 (December). Real-Time Control of CMU Direct-Drive Arm II Using Customized Inverse Dynamics. *Proc. of 23rd Conf. on Decision and Control.* Las Vegas, Nevada: pp. 1345-1352.

Kanade, T., and Schmitz, D. 1985 (June 19-21). Development of the CMU Direct Drive Arm II. *Proc. American Control Conf..* Boston: pp. 703-709.

Kazerooni, H. 1985 (Feb.). A Robust Design Method for Impedance Control of Constrained Dynamic Systems. Ph.D. thesis, MIT, Mechanical Engineering Department.

Kazerooni, H., Houpt, P. K., and Sheridan, T. B. 1986a (April 7-10). The fundamental concepts of robust compliant motion for robot manipulators.

Proc. IEEE Int. Conf. Robotics and Automation. San Francisco: pp. 418-427.

Kazerooni, H., Sheridan, T. B., Houpt, P. K. 1986b (June). Robust compliant motion for manipulators, Part I: The fundamental concepts of compliant motion. *IEEE Journal of Robotics and Automation.* RA-2: 83-92.

Kazerooni, H., Houpt, P. K., and Sheridan, T. B. 1986c (June). Robust compliant motion for manipulators, Part II: Design method. *IEEE Journal of Robotics and Automation.* RA-2: 93-105.

Khalil, W., Gautier, M., and Kleinfinger, J. F. 1986. Automatic Generation of Identification Models of Robots. *International Journal of Robotics and Automation.* 1(1): 2-6.

Khatib, O. 1983 (Dec 15-20). Dynamic control of manipulators in operational space. *Sixth IFTOMM Congress on Theory of Machines and Mechanisms.* New Delhi: pp. 1128-1131.

Khatib, O. 1987. A unified approach for motion and force control of robot manipulators: the operational space formulation. *IEEE J. Robotics and Automation.* RA-3: 43-53.

Khatib, O. and Burdick, J. 1986 (April 7-10). Motion and force control of robot manipulators. *Proc. IEEE Int. Conf. Robotics and Automation.* San Francisco: pp. 1381-1386.

Khosla, P. K. 1986 (August). Real-Time Control and Identification of Direct-Drive Manipulators. Ph.D. thesis, Carnegie-Mellon Univ., Department of Electrical and Computer Engineering.

Khosla, P. K. 1987 (March 31-April 3). Choosing sampling rates for robot control. *Proc. IEEE Int. Conf. Robotics and Automation.* Raleigh, NC: pp. 69-174.

Khosla, P. K., and Kanade, T. 1985 (Dec. 11-13). Parameter identification of robot dynamics. *Proc. 24th Conf. Decision and Control.* Fort Lauderdale, Florida: pp. 1754-1760.

Khosla, P. K., and Kanade, T. 1986 (April 7-10). Real-time implementation and evaluation of model-based controls on CMU DD Arm II. *Proc. IEEE Conf. on Robotics and Automation.* San Francisco: pp. 1546-1555.

Kumar, A., and Waldron, K. J. 1981. Numerical plotting of surfaces of positioning accuracy of manipulators. *Mechanism and Machine Theory.* 16: 361-368.

Kuwahara, H., Ono, Y., Nikaido, M., and Matsumoto, T. 1985 (June 19-21). A precision direct-drive robot arm. *Proc. American Control Conf..* Boston: pp. 722-727.

Kuwahara, H., Ono, Y., Nikaido, M., and Matsumoto, T. 1986. Development of high torque/precision servo system for direct-drive manipulator. *Robotics Research: The Third International Symposium.*, ed. O. Faugeras and G. Giralt. Cambridge, Mass.: MIT Press, pp. 297-306.

Lau, K., Hocken, R., and Haynes, L. 1985. Robot performance measurements using automatic laser tracking techniques. *Robotics & Computer-Integrated Manufacturing.* 2: 227-236.

Leahy, M. B., Valavanis, K. P., and Saridis, G. N. 1986 (April 7-10). The effects of dynamic models on robot control. *Proc. IEEE Conf. on Robotics and Automation.* San Francisco: pp. 49-54.

Leahy, M. B., Valavanis, K. P. 1987 (Mar. 31–Apr. 2). Compensation of unmodelled PUMA manipulator dynamics. *Proc. IEEE Int. Conf. Robotics and Automation.* Raleigh, N.C.: pp. 151-156.

Lee, K. 1983 (December). Shape Optimization of Assemblies Using Geometric Properties. Ph.D. thesis, MIT, Mechanical Engineering Department.

Lehtomaki, N. A. 1981 (May). Practical Robustness Measures in Multivariable Control System Analysis. Ph.D. thesis, MIT.

Liégeois, A., Fournier, A. and Aldon, M. 1980 (August 13-15). Model Reference Control of High-Velocity Industrial Robots. *Proc. Joint Automatic Control Conf..* San Francisco, CA:.

Lipkin, H., and Duffy, J. 1986 (Sept. 9-12). Invariant kinestatic filtering. *Preprints 6th CISM-IFToMM Symp. on Theory and Practice of Robots and Manipulators*. Cracow, Poland: pp. 96-103.

Ljung, L. and Soderstrom, T. 1983. *Theory and Practice of Recursive Identification.* Cambridge, Ma.: MIT Press.

Luh, J. Y. S., Fisher, W. D, and Paul, R. 1983 (Feb.). Joint torque control by a direct feedback for industrial robots. *IEEE Trans. Automatic Control.* AC-28: 153-160.

Luh, J. Y. S., Walker, M., and Paul, R. 1980a. On-line computational scheme for mechanical manipulators. *J. Dynamic Systems, Meas., Control.* 102: 69-76.

Luh, J. Y. S., Walker, M., and Paul, R. 1980b. Resolved-acceleration control of mechanical manipulators. *IEEE Trans. Auto. Contr..* AC-25: 468-474.

Markiewicz, B. 1973 (March 15). Analysis of the Computed Torque Drive Method and Comparison With Conventional Position Servo for a Computer-Controlled Manipulator. Pasadena, CA: Jet Propulsion Laboratory.

Marquardt, D. W., and Snee, R. D. 1975. Ridge regression in practice. *Amer. Statistician.* 29: 3-20.

Mason, M. T. 1981. Compliance and Force Control for Computer Controlled Manipulators. *IEEE Transactions on Systems, Man and Cybernetics.* SMC-11(6): 418–432.

Mathlab Group 1983. MACSYMA Reference Manual. Version 10. Cambridge, MA.: Massachusetts Institute of Technology, Laboratory for Computer Science.

Mayeda, H., Osuka, K., and Kangawa, A. 1984 (July 2-6). A new identification method for serial manipulator arms. *Preprints IFAC 9th World Congress, Vol. VI.* Budapest: pp. 74-79.

Mita, T., and Kato, E. 1985 (Dec. 11-13). Iterative Control and its Application to Motion Control of Robot Arm – A Direct Approach to Servo-Problems. *Proc. 24th Conf. on Decision and Control.* Fort Lauderdale, Florida: pp. 1393-1398.

Mooring, B. W., and Tang, G. R. 1984. An improved method for identifying the kinematic parameters in a six-axis robot. *ASME Proc. Int. Computers in Engineering Conf.*. Las Vegas: pp. 79-84.

Morita, A. 1986 (February 27). A Study of Learning Controllers For Robot Manipulators With Sparse Data. M.S. thesis, Massachusetts Institute of Technology, Mechanical Engineering.

Mukerjee, A. 1984 (November). Adaptation in biological sensory-motor systems: A model for robotic control. *Proc. SPIE Conf. on Intelligent Robots and Computer Vision, Vol. 521.* Cambridge, Ma.:.

Mukerjee, A. 1986 (Sept.). Self-Calibration Strategies for Robot Manipulators. TR 193. Rochester, N.Y.: Dept. Computer Science, Univ. Rochester.

Mukerjee, A., and Ballard, D. H. 1985 (Mar. 25-28). Self-calibration in robot manipulators. *Proc. IEEE Conf. Robotics and Automation.* St. Louis: pp. 1050-1057.

Narasimhan, S., Siegel, D. M., Biggers, K., and Gerpheide, G. 1986 (April 7-10). Implementation of control methodologies on the computational architecture of the Utah/MIT hand. *Proc. IEEE Int. Conf. Robotics and Automation.* San Francisco: pp. 1884-1889.

Narendra, K. S., and Kudva, P. 1974 (Nov). Stable adaptive schemes for system identification and control - Part II. *IEEE Trans. on Systems, Man, and Cybernetics.* SMC-4(6): 552-560.

Narendra, K. S., and Valavani, L. S. 1977 (Feb). Stable adaptive observers and controllers. *IEEE Proceedings.* 64: 1198-1208.

Olsen, H. B., and Bekey, G. A. 1985 (Mar. 25-28). Identification of parameters in models of robots with rotary joints. *Proc. IEEE Conf. Robotics and Automation.* St. Louis: pp. 1045-1050.

Olsen, H. B., and Bekey, G. A. 1986 (April 7-10). Identification of robot dynamics. *Proc. IEEE Int. Conf. Robotics and Automation.* San Francisco: pp. 1004-1010.

Paul, R. Sept. 1972. Modeling,trajectory calculation and servoing of a computer controlled arm. AIM-177. Stanford University Artificial Intelligence Laboratory.

Paul, R. 1981. *Robot Manipulators: Mathematics, Programming, and Control.* Cambridge, Mass.: MIT Press.

Pawletko, J. P. and Chai, H. D. 1973. Linear Step Motors. *Theory and Applications of Step Motors.*, ed. B.C. Kuo. West Publishing Co., pp. 316-326.

Puskorius, G. V., and Feldkamp, L.A. 1987 (March 31-April 3). Global calibration of a robot/vision system. *Proc. IEEE Int. Conf. Robotics and Automation.* Raleigh, NC: pp. 190-195.

Raibert, M. H., and Craig, J. J 1981 (June). Hybrid position/force control of manipulators. *ASME J. of Dynamic Systems, Measurement, and Control.* 102: 126-133.

Ramirez, R. 1984 (Feb.). Design of High Speed Graphite Composite Robot Arm. MS thesis, MIT, Department of Mechanical Engineering.

Roberge, J. K 1975. *Operational Amplifiers: Theory and Practice.* New York: John Wiley and Sons, Inc..

Roberts, R. K 1984 (Dec.). The Compliance of End Effector Force Sensors for Robot Manipulator Control. Ph.D. thesis, Purdue Univ., School of Electrical Engineering.

Roberts, R. K., Paul, R., and Hillberry, B. M. 1985 (Mar. 25-28). The effect of wrist force sensor stiffness on the control of robot manipulators. *Proc. IEEE Conf. Robotics and Automation.* St. Louis: pp. 269-274.

Salisbury, J. K. 1980 (Dec.). Active stiffness control of a manipulator in Cartesian coordinates. *Proc. 19th IEEE CDC.* pp. 95-100.

Salisbury, J. K. 1985. Kinematic and force analysis of articulated hands. *Robot Hands and the Mechanics of Manipulation.*, ed. Mason, M.T., and Salisbury, J.K.. Cambridge, MA: MIT Press, pp. 2-167.

Samson, C. 1983 (Dec. 14-16). Robust nonlinear control of robotic manipulators. *Proc. 22nd IEEE Conf. Decision and Control.* San Antonio:.

Shapiro, R. 1978. Direct linear transformation method for three-dimensional cinematography. *Research Quarterly.* 49: 197-205.

Shih, S. 1985 (March). Reduced-order model-reference adaptive system identification of large scale systems with discrete adaptation laws. Ph.D. thesis, MIT, Department of Electrical Engineering and Computer Science.

Shin, K. G., and Lee, C.-P. 1985. Compliant control of robotic manipulators with resolved acceleration. *Proc. 24th IEEE CDC.* Ft. Lauderdale: pp. 350-357.

Siegel, D. M., Garabieta, I., and Hollerbach, J. M. 1986 (April 7-10). An integrated tactile and thermal sensor. *Proc. IEEE Int. Conf. Robotics and Automation.* San Francisco: pp. 1286-1291.

Slotine, J.-J. E. 1985. The robust control of robot manipulators. *Int. J. Robotics Research.* 4(2): 49-64.

Slotine, J.-J. E., and Li, W. 1987 (Mar 31 - April 3). Adaptive manipulator control: a case study. *Proc. IEEE Int. Conf. Robotics and Automation.* Raleigh: pp. 1392-1400.

Spong, M. W., Thorp, J. S., and Kleinwaks, J. M. 1984 (Dec. 12-14). The control of robot manipulators with bounded input. Part II: robustness and disturbance rejection. *Proc. 23rd IEEE Conf. Decision and Control.* Las Vegas: pp. 1047-1052.

Stone, H. W., Sanderson, A. C., and Neuman, C. P. 1986 (April 7-10). Arm signature identification. *Proc. IEEE Int. Conf. Robotics and Automation.* San Francisco: pp. 41-48.

Stone, H. W., and Sanderson, A. C. 1987 (March 31-April 3). A prototype arm signature identification system. *Proc. IEEE Int. Conf. Robotics and Automation.* Raleigh, NC: pp. 175-182.

Sugimoto, K., and Okada, T. 1985. Compensation of positioning errors caused by geometric deviations in robot system. *Robotics Research: The Second International Symposium.*, ed. H. Hanafusa and H. Inoue. Cambridge, Mass.: MIT Press, pp. 231-236.

References

Symon, K. R. 1971. *Mechanics.* Reading, Mass.: Addison-Wesley.

Togai, M., and Yamano, O. 1985 (Dec. 11-13). Analysis and Design of an Optimal Learning Control Scheme for Industrial Robots: A Discrete Time Approach. *Proc. 24th Conf. on Decision and Control.* Fort Lauderdale, Florida: pp. 1399-1404.

Togai, M., and Yamano, O. 1986 (April 7-10). Learning Control and Its Optimality: Analysis and Its Application to Controlling Industrial Robots. *Proc. 1986 IEEE International Conference on Robotics and Automation.* San Francisco, CA: pp. 248-253.

Tsai, L.-W., and Morgan, A. P. 1985. Solving the kinematics of the most general six- and five-degree-of-freedom manipulators by continuation methods. *ASME J. Mechanisms, Transmissions, and Automation in Design.* 107: 189-200.

Uchiyama, M. 1978. Formation of High-Speed Motion Pattern of a Mechanical Arm by Trial. *Trans. of Society of Instrument and Control Engineers (Japan).* 19(5): 706–712.

Veitschegger, W. K., and Wu, C.-H. 1987 (March 31-April 3). A method for calibrating and compensating robot kinematic errors. *Proc. IEEE Int. Conf. Robotics and Automation.* Raleigh, NC: pp. 39-44.

Wang, S. H. 1984 (October). Computed Reference Error Adjustment Technique (CREATE) For The Control of Robot Manipulators. *22nd Annual Allerton Conf. on Communication, Control, and Computing..*

Wang, S. H., and Horowitz, I. 1985 (March). CREATE - A New Adaptive Technique. *Proc. of the Nineteenth Annual Conf. on Information Sciences and Systems.* pp. 620-622.

Whitney, D. E. 1972. The mathematics of coordinated control of prosthetic arms and manipulators. *ASME J. Dynamic Systems, Meas., Control.*: 303-309.

Whitney, D. E. 1987. Historical perspective and state of the art in robot force control. *Int. J. Robotics Research.* 6(1): 3-14.

Whitney, D. E., Lozinski, C. A., and Rourke, J. M. 1986. Industrial robot forward calibration method and results. *J. Dynamic Systems, Meas., Control.* 108: 1-8.

Wlassich, J. J. 1986 (Feb.). Nonlinear Force Feedback Impedance Control. MS thesis, MIT, Dept. of Mechanical Engineering.

Wu, C. H. and Paul, R. 1980 (Dec. 10-12). Manipulator compliance based on joint torque control. *Proc. IEEE Conf. Decision and Control.* Alburquerque, NM: pp. 88-94.

Youcef-Toumi, K. 1985 (May). Analysis, Design, and Control of Direct-Drive Manipulators. Ph.D. thesis, MIT, Dept. of Mechanical Engineering.

Index

Actuator command, 122-124
    command error, 123, 132
Actuator
    dynamics, 120, 128
    noise, 119-120, 128, 135
    See also Motor
Adaptation, 89
Adaptive control, 113, 136
    adaptive feedforward, 120
    adaptive feedback, 120
Adaptive observer, 155
Adaptive processes, 113
AdeptOne, 38-39
Aluminum surface, 145, 160
    force step response on, 160
    stiffness estimation of, 145
Amplifier, 10, 33, 44, 157, 110, 161, 197
    bandwidth, 44, 157
    commutation, 10, 44, 110, 159, 197
    deadzone, 44, 99, 159, 161, 189
    nonlinearity, 110, 197
    PWM, 44
    slew rate, 33
Analog servo, 46, 101, 104, 111
Angular velocity vector, 13, 56, 69

Backlash, 10, 33, 34, 37, 41, 51, 94, 140

Bandwidth, 33, 34, 44, 128, 156, 161, 166, 189
Bias errors, 81, 145, 151, 161, 189

Calibration (kinematic)
    cube, 52
    frame, 51, 61
Cartesian-based force control
    hybrid position/force, 25, 28, 168, 171, 176, 191
    hybrid position/force feedforward, 8
    impedance, 140, 141, 170
    operational space, 168, 208
    PD, 7
    resolved acceleration, 28, 168, 181, 185, 193, 208
    stiffness, 25, 140, 141, 168, 174, 184, 193
Cartesian-based position control
    feedforward controller, 25
    resolved acceleration, 26, 168
    PD, 25
Center of mass (gravity), 13, 65, 69, 71-72, 75-77, 79, 84
Centripetal torques, 22
Closed form dynamics, 93, 204
CMU Direct Drive Arm II, 39, 89
Cogging, 10, 44, 159, 161
Command disturbance, 123

Command following, 161
Commutation, 10, 44, 110, 159, 197
Compliance, 81
Compliant coverings, 148
Computed torque control, 16, 19, 87, 101, 103, 107
Continuous time, 124
Control, 4
Control algorithms: see
    Cartesian-based force control
    Cartesian-based position control
    Joint-based force control
    Joint-based position control
Convergence, 130, 136
    rate, 118-119
Convolution, 131
Coordinate frames, 52-54, 67, 90
Coriolis torques, 22, 87
Counter-balancing, 89
Currents, 94

Damping, 128, 131
DDArm: see MIT Serial Link Direct Drive Arm
Deadzone, 44, 99, 159, 161, 189
Delay, 17, 131, 132
Denavit-Hartenberg parameters, 11, 41, 52, 61
Differential kinematics, 11, 55
Differentiating filter, 73, 127
Digital servo, 46, 102
Direct drive arm, 2, 49-64, 122
    AdeptOne, 38-39
    CMU Direct Drive Arm II, 39, 89
    five-bar-linkage design, 38
    MIT Serial Link Direct Drive Arm (DDArm), 2, 38, 44-50, 119, 124, 136
    YEWBOT, 38, 40
Direct-drive motor, 10, 37-38, 43-44, 99, 197-198
    ISI, 38, 43-44, 99, 110, 157, 196
    variable reluctance, 37-38, 40, 197
Direct dynamics, 7, 122
Direct kinematics, 5, 11, 54, 115
Direct linear transformation (DLT), 52
Discrete time, 124, 131
Discretization, 135
Disturbance, 119-120, 123, 127-128, 135-136
    rejection, 161
Dominant pole, 157-159, 188
Dynamic estimation, 76, 79
Dynamic instability, 23, 139-166
Dynamic model, 19-21, 29, 114
Dynamics
    closed form, 93, 204
    direct, 7, 122
    inverse, 8, 15, 18, 22, 27, 102, 123, 176
    Lagrangian, 88
    linearized, 176
    Newton-Euler, 13, 67-71, 90-92
    See also Inertial parameters

Eccentric cam, 165, 188-189
Eigen-structure assignment, 139
Eigenvalue problem, 59
Eigenvalues, 118, 132, 133, 175
Ellipsoid of inertia, 59
Endpoint variation, 56
Environment, 139, 142
    environment stiffness, 148

Feedback control: see control algorithms

Index

Feedforward command, 118-119, 122, 127, 130
    error, 123, 130-135
    initialization, 120, 122, 124, 128
    memory, 122
    modification, 123
    update, 124
    torques, 124
Feedforward control, 7, 18
Feedforward controller, 16, 18, 102, 104, 106
Filter, 120, 136
    differentiating, 73, 127
    low-pass, 127, 157-160, 166
Fingers, 148
Finite learning interval convergence test, 133
Finite time convergence, 135
Five-bar-linkage design, 38
Fixed point theory, 117, 130
Flexible link dynamics, 120
Flexible modes, 144
Force control, 22-30, 139-166, 167-194
Force control algorithms: see
    Cartesian-based force control
    Dynamic instability
    Joint-based force control
    Kinematic instability
Force resolution, 156
Force step response, 160
Force/torque sensor, 9, 14, 15, 43, 44, 65, 70, 73, 79, 89, 145, 156, 188
Force vector, 6, 13, 25, 69, 71, 91, 168
Forward kinematics: see Direct kinematics
Friction, 10, 33, 34, 37, 41, 89, 94, 101, 102, 107, 110, 111, 140, 156, 180, 181

Gain, 135
Gears, 10, 33-35, 37, 38, 40, 43, 50, 103, 110, 156, 197
    backlash, 10, 33, 34, 37, 41, 51, 94, 140
    flexibility, 10, 33
Global stability, 175
Gravity
    torques, 22, 75, 84, 104
    vector, 69

Hayati parameters, 53
Height gauge, 63
High gain feedback, 143, 144
Hybrid analog/digital controller, 101, 104, 111
Hybrid control, 170
    See Hybrid position/force control
Hybrid position/force feedforward controller, 8
Hybrid position/force control, 25, 28, 168, 171, 176, 191

Impedance control, 140, 141, 170
Impulse response, 131-133
Independent joint PD control, 128
    See also PD control
Inertia matrix, 22, 87, 128, 176, 181, 193
Inertial parameters, 65, 87
    CAD-modelled, 42, 94, 97-99
    center of mass (gravity), 13, 65, 69, 71-72, 75-77, 79, 84
    link estimation, 14-16, 87-100
    load estimation, 12-14, 65-85
    mass, 13, 65, 69
    mass moment, 13, 71

moments of inertia, 13, 65, 69-70, 84-85
principal axes, 69, 72
principal moments, 72
unidentifiability of, 89, 93, 97
Infrared light-emitting diodes, 51
Instability (in force control)
dynamic instability, 139-166
kinematic instability, 167-194
Integral load estimation equations, 201
Internal model, 113
Inverse dynamics, 8, 15, 18, 22, 27, 102, 123, 176
Inverse kinematics, 6, 115-116
Inverse model, 113-119, 131-136
Iterative learning, 114
Iterative least squares (kinematics), 64

Jacobian, 168
  inverse, 6, 25, 118, 170, 175, 193
  matrix, 6, 11, 25, 55-56, 118
  transpose, 6, 28-29, 170, 175, 193
Joint-based force control
  PD, 7
  Stiffness, 168
Joint-based position control
  computed torque controller, 16, 19, 87, 101, 103, 107
  feedforward controller, 16, 18, 102, 104, 106
  PD, 16, 102, 104, 106
Joint compliance dynamics, 120
Joint coordinates, 5
Joint torque, 6, 92, 94, 97
  control, 34, 37, 44, 111, 156, 188, 197

sensing, 34, 67, 88-89, 110, 156, 159

Kinematic calibration
  parametric, 11, 49, 87
  nonparametric, 116
Kinematic instability, 23, 167-195
Kinematic learning, 114-115
  convergence, 117
Kinematic parameters: see Denavit-Hartenberg parameters
Kinematic transformation, 167, 170
Kinematics
  direct, 5, 11, 54, 115
  inverse, 6, 115
  linearized, 11, 55, 117, 119
  redundancy, 6, 115
  singularity, 6, 115

Lagrangian dynamics, 88
Laplace transform, 122-123
  error, 124
Laser tracking system, 50
Lateral-effect photodiodes, 52
Learning algorithm, 114, 127-129
  convergence, 118, 135
  derivation, 136
  failure, 135
  performance, 135-136
Learning efficiency, 113
Learning from practice, 136
Learning interval, 131-135
Learning of non-repetitive tasks, 136
Learning operator, 113, 117-119, 127-136
  bandwidth, 135
  convergence, 119
  efficacy, 135

Index

impulse response, 131, 133
Learning performance, 129, 136
Least squares, 65, 72-73, 92, 150-151
Levenberg-Marquardt algorithm, 62, 64
Linear learning operator, 119
Linearized dynamics, 176
Linearized kinematics, 11, 55, 117, 119
Link estimation, 14-16, 87-100
   by adaptation, 89
   unidentifiability, 89, 93, 97
Load estimation, 12-14, 65-86
   dynamic estimation, 76, 79
   orientation, 69, 84
   static identification, 75, 76
Local stability, 175-176
Loop gain, 156
Low-pass filter, 127, 157-160, 166

MACSYMA, 93
Mass, 13, 65, 69
Mass moment, 13, 71
Measurement noise, 128, 135
Memory, 118
Minimum phase, 131
MIT Serial Link Direct Drive Arm (DDArm), 2, 38, 44-50, 119, 124, 136
Model
   accuracy, 118
   approximate, 124
   bad, 124
   errors, 87, 114, 116, 120, 123, 129, 136, 143, 206
   imperfect, 116
   perfect, 114, 119, 124
   usefulness, 1, 110, 136, 193, 195

Model (parametric)
   kinematic parameters, 11, 49-64, 115
   link inertial parameters, 14, 87-100
   load inertial parameters, 12, 65-85
   motor, 10, 44
Model-based (trajectory) learning, 113-138
   command refinement, 136
Moments of inertia, 13, 65, 69-70, 84-85
Motion tracking systems, 49-51
   Laser tracking system, 50
   Selspot system, 50
   Watsmart system, 12, 51, 64
Motor, 10-11, 31-48
   cogging, 10, 44, 159, 161
   currents, 156
   model, 10, 44
   torque, 33, 37, 44, 110, 197
   See also Direct drive motor

Newton-Euler equations, 13, 69, 88, 89
   for load, 13, 67
   for links, 90
Noise, 71, 75, 81, 119-120, 123, 127
Nonlinearity cancellation, 19
Nonlinear convergence criteria, 129
Nyquist criterion, 206

Open-loop force control, 156
Operational space control, 168, 208
Optimal filter, 83
Optimization, 113
Orientation, 69, 84

Parallel axis theorem, 70, 72
Parameter estimation, 9

PD control: see
    Cartesian-based force control
    Cartesian-based position control
    Joint-based force control
    Joint-based position control
Performance measures, 161
Persistently exciting trajectories, 100
Perturbation, 117
PID control, 101
Planning, 4
Plant dynamics, 120
Polar manipulator, 173, 180
Position control: see
    Cartesian-based position control
    Joint-based position control
Position sensors, 127
    See also Resolver
Positive feedback, 173
Practice, 113-127, 135
Precision points, 49
Principal axes, 69, 72
Principal moments, 72
PUMA, 73, 76, 83, 89, 128

Recognition, 66
Region of instability, 180
Resolved acceleration
    position control, 26, 168
    force control, 28, 168, 181, 185, 193, 208
Resolver, 43, 94, 110, 155, 197
Resonant frequency, 157
Ridge regression, 93, 95, 98
Rigid body dynamics: see Dynamics
Robot hand, 66
Robot model: see Model

Root locus, 177
Rotary manipulator, 171, 176
Rotor inertia, 33-35, 37, 44, 89, 157
Rubber surface, 145
    stiffness estimation of, 145
RMS error, 62

Sampling frequency, 73, 79, 107-111, 131, 135, 151, 189
Screw coordinates, 55
Selection matrix, 28, 168, 172, 189
Selspot system, 50
Sensor noise, 119-120, 127, 135
Sensor stiffness, 148
Sensors
    force/torque, 9, 14, 15, 43, 44, 65, 70, 73, 79, 89, 145, 156, 188
    resolver, 43, 94, 110, 155, 197
    tachometer, 41, 79, 94, 110
    Watsmart system, 12, 51, 64
Servo: see
    Analog servo
    Digital servo
Signal to noise ratio, 76
Similarity transform, 182
Simplified models, 114, 127, 128
Singular value decomposition, 93, 98
Singularity, 6, 115, 173
Sinusoidal trajectory, 161, 189
Special test movements, 67, 77, 79, 88
Stability, 119
Stanford Arm, 181
Static identification, 75, 76
Statics, 6
Steady state accuracy, 159, 161
Step response, 160, 189
Stereo camera system, 49

Index

Stiffness control, 25, 140, 141, 168, 174, 184, 193
Stiffness matrix, 168, 174
Straight line estimation, 57
Strain gauges, 156
System identification, 9, 113

Tachometer, 41, 79, 94, 110
Tactile sensors, 148
Task coordinates, 4
Theodolites, 49
Toeplitz matrix, 132-133
Torque: see
    Joint torque
    Motor torque
Torque vector, 13, 69, 75, 91
Trajectory
    fifth order polynomial, 74, 83, 94, 105, 124
    persistently exciting, 100
    seventh order polynomial, 128
    sinusoidal, 161, 189
    special test, 67, 77, 79, 88
Trajectory learning, 20, 113-138
    algorithm, 119-124
    convergence, 114, 119, 129, 131
    convergence test, 133
    implementation, 124
    interval, 130
    performance, 120
Transfer function impulse response, 132
Triangular path, 183, 185

Ultrasonic range sensor, 49
Unidentifiability (link inertial parameters), 89, 93, 97
Unmodeled dynamics, 81, 127, 143
    in learning, 127
    in load estimation, 81
    in force control, 143

Variance accounted for, 62
Velocity sensors, 127
    See also Tachometer
Verification, 66
Vibration, 83
Vision system, 115

Watsmart system, 12, 51, 64
Wrench, 91
    transmission matrix, 91
Wrist force sensing, 156, 159
    See also Force/torque sensor

YEWBOT, 38, 40

The MIT Press Series in Artificial Intelligence
Edited by Patrick Henry Winston and Michael Brady

Artificial Intelligence: An MIT Perspective, Volume I: Expert Problem Solving, Natural Language Understanding, Intelligent Computer Coaches, Representation and Learning edited by Patrick Henry Winston and Richard Henry Brown, 1979

Artificial Intelligence: An MIT Perspective, Volume II: Understanding Vision, Manipulation, Computer Design, Symbol Manipulation edited by Patrick Henry Winston and Richard Henry Brown, 1979

NETL: A System for Representing and Using Real-World Knowledge by Scott Fahlman, 1979

The Interpretation of Visual Motion by Shimon Ullman, 1979

A Theory of Syntactic Recognition for Natural Language by Mitchell P. Marcus, 1980

Turtle Geometry: The Computer as a Medium for Exploring Mathematics by Harold Abelson and Andrea diSessa, 1981

From Images to Surfaces: A Computational Study of the Human Early Visual System by William Eric Leifur Grimson, 1981

Robot Manipulators: Mathematics, Programming and Control by Richard P. Paul, 1981

Computational Models of Discourse edited by Michael Brady and Robert C. Berwick, 1982

Robot Motion: Planning and Control edited by Michael Brady, John M. Hollerbach, Timothy Johnson, Tomás Lozano-Pérez, and Matthew T. Mason, 1982

In-Depth Understanding: A Computer Model of Integrated Processing for Narrative Comprehension by Michael G. Dyer, 1983

Robotics Research: The First International Symposium edited by Michael Brady and Richard Paul, 1984

Robotics Research: The Second International Symposium edited by Hideo Hanafusa and Hirochika Inoue, 1985

Robot Hands and the Mechanics of Manipulation by Matthew T. Mason and J. Kenneth Salisbury, Jr., 1985

The Acquisition of Syntactic Knowledge by Robert C. Berwick, 1985

The Connection Machine by W. Daniel Hillis, 1985

Legged Robots that Balance by Marc H. Raibert, 1986

Robotics Research: The Third International Symposium edited by O. D. Faugeras and Georges Giralt, 1986

Machine Interpretation of Line Drawings by Kokichi Sugihara, 1986

ACTORS: A Model of Concurrent Computation in Distributed Systems by Gul A. Agha, 1986

Knowledge-Based Tutoring: The GUIDON Program by William Clancey, 1987

AI in the 1980s and Beyond: An MIT Survey edited by W. Eric L. Grimson and Ramesh S. Patil, 1987

Visual Reconstruction by Andrew Blake and Andrew Zisserman, 1987

Reasoning about Change: Time and Causation from the Standpoint of Artificial Intelligence by Yoav Shoham, 1988

Model-Based Control of a Robot Manipulator by Chae H. An, Christopher G. Atkeson, and John M. Hollerbach, 1988

The MIT Press, with Peter Denning, general consulting editor, and Brian Randell, European consulting editor, publishes computer science books in the following series:

ACM Doctoral Dissertation Award and Distinguished Dissertation Series

Artificial Intelligence, Patrick Winston and Michael Brady, editors

Charles Babbage Institute Reprint Series for the History of Computing, Martin Campbell-Kelly, editor

Computer Systems, Herb Schwetman, editor

Exploring with Logo, E. Paul Goldenberg, editor

Foundations of Computing, Michael Garey and Albert Meyer, editors

History of Computing, I. Bernard Cohen and William Aspray, editors

Information Systems, Michael Lesk, editor

Logic Programming, Ehud Shapiro, editor; Fernando Pereira, Koichi Furukawa, and D. H. D. Warren, associate editors

The MIT Electrical Engineering and Computer Science Series

Scientific Computation, Dennis Gannon, editor